Slow Tourism

ASPECTS OF TOURISM

Series Editors: Chris Cooper *(Oxford Brookes University, UK)*, C. Michael Hall *(University of Canterbury, New Zealand)* and Dallen J. Timothy *(Arizona State University, USA)*

Aspects of Tourism is an innovative, multifaceted series, which comprises authoritative reference handbooks on global tourism regions, research volumes, texts and monographs. It is designed to provide readers with the latest thinking on tourism worldwide and push back the frontiers of tourism knowledge. The volumes are authoritative, readable and user-friendly, providing accessible sources for further research. Books in the series are commissioned to probe the relationship between tourism and cognate subject areas such as strategy, development, retailing, sport and environmental studies.

Full details of all the books in this series and of all our other publications can be found on http://www.channelviewpublications.com, or by writing to Channel View Publications, St Nicholas House, 31–34 High Street, Bristol BS1 2AW, UK.

ASPECTS OF TOURISM
Series Editors: Chris Cooper *(Oxford Brookes University, UK)*, C. Michael Hall *(University of Canterbury, New Zealand)* and Dallen J. Timothy *(Arizona State University, USA)*

Slow Tourism

Experiences and Mobilities

Edited by
Simone Fullagar, Kevin Markwell and Erica Wilson

CHANNEL VIEW PUBLICATIONS
Bristol • Buffalo • Toronto

Library of Congress Cataloging in Publication Data
A catalog record for this book is available from the Library of Congress.
Slow Tourism: Experiences and Mobilities/Edited by Simone Fullagar, Kevin Markwell and Erica Wilson.
Aspects of Tourism: 54
Includes bibliographical references.
1. Tourism--Psychological aspects. 2. Social movements. I. Fullagar, Simone. II. Markwell, Kevin. III. Wilson, Erica.
G155.A1S5625 2012
306.4'819–dc23 2011048974

British Library Cataloguing in Publication Data
A catalogue entry for this book is available from the British Library.

ISBN-13: 978-1-84541-281-4 (hbk)
ISBN-13: 978-1-84541-280-7 (pbk)

Channel View Publications
UK: St Nicholas House, 31–34 High Street, Bristol BS1 2AW, UK.
USA: UTP, 2250 Military Road, Tonawanda, NY 14150, USA.
Canada: UTP, 5201 Dufferin Street, North York, Ontario M3H 5T8, Canada.

The policy of Multilingual Matters/Channel View Publications is to use papers that are natural, renewable and recyclable products, made from wood grown in sustainable forests. In the manufacturing process of our books, and to further support our policy, preference is given to printers that have FSC and PEFC Chain of Custody certification. The FSC and/or PEFC logos will appear on those books where full certification has been granted to the printer concerned.

Typeset by Techset Composition Ltd., Salisbury, UK.
Printed and bound in Great Britain by Short Run Press Ltd.

Contents

Acknowledgements

The authors would like first and foremost to thank all of the contributors to this book, who have provided such a diversity of perspectives on this fascinating and complex thing called 'slow tourism'. They have been a pleasure to work with, and we appreciate their willingness to work quickly (ironically) to meet the tight timelines required for publication. Thank you also to Elinor Robertson, and the team at Channel View, for supporting the development of the slow mobilities book from the initial idea to final publication. We would further like to acknowledge the research assistance of James Catlin, and the editorial work on final drafts by Jean Martinez. Every reasonable effort has been made to locate, contact and acknowledge copyright owners. In relation to Chapter 12 we thank Dan Heringa Photography for permission to use the image: Somewhere down there is your SOUL and Natural Resources Canada for permission to use the map of Yukon courtesy of the Atlas of Canada.

Simone would like to thank Adele Pavlidis for her research assistance on the cycling project and the women participants who gave their time to the research as they inspired the slow travel project. She would also like to acknowledge the Griffith Centre for Cultural Research, Griffith University, for seed funding that enabled the slow travel research and the Griffith Business School for providing research leave. Finally, Simone would like to thank her partner Gill for her support through the process and for embracing the spirit of slow travel and living.

Both Kevin and Erica acknowledge the support of the School of Tourism and Hospitality Management at Southern Cross University for providing such an amiable and creative research culture which allows time for slow when needed, and for recognising the importance of slow living outside of work. Kevin thanks his partner Steve Harrison for his sustained and good humoured support. Erica would particularly like to thank her partner Noah, and her children Maya and Solomon, for continually reminding her of what is most important in life, and for helping her to put slow living in practice.

Entrance to Royal Botanical Gardens Cranbourne, Victoria, Australia
(E. Wilson)

Contributors

Editors

Simone Fullagar is an interdisciplinary sociologist who has published widely across the areas of health, leisure and tourism, using post-structuralist and feminist perspectives. She is Associate Professor in the Department of Tourism, Leisure, Hotel and Sport Management at Griffith University, Queensland, Australia. Simone has completed several large ARC grant projects involving qualitative research into women's recovery from depression and the sociocultural context of youth suicide. She has also undertaken research on the government of family leisure, healthy lifestyles and the politics of risk as well as new work on the rise of green lifestyles and slow travel.

Kevin Markwell is a cultural geographer whose research interests focus on the contributions that nature-based/eco/wildlife tourism make as tools for nature conservation. In addition he is interested in understanding the ways by which leisure practices can act as vehicles for the construction of gay identities. He is Associate Professor in the School of Tourism and Hospitality Management at Southern Cross University, NSW, Australia and is also Adjunct Associate Professor in Tourism at Divine Word University, Madang, PNG. His most recent book is *Snake Bitten, Eric Worrell and the Australian Reptile Park*, co-authored with colleague, Nancy Cushing.

Erica Wilson is Senior Lecturer in the School of Tourism and Hospitality Management at Southern Cross University. Her doctoral thesis was an interpretive exploration into the constraints faced by Australian solo women travellers. Erica also holds a postgraduate diploma in environmental studies and an honours degree in tourism administration. Erica teaches in the areas of sustainable tourism and special interest tourism, and her research publications reflect her scholarly interests in women's travel and adventure, work–life balance, sustainable tourism and critical approaches to tourism research. When she is not parenting or working, Erica enjoys living in a solar-powered, rammed earth home that she and her partner built – very slowly – on their property near Nimbin.

Authors

Paulo de Abreu e Lima is the founder of ESTILOGOURMAND, a quality food research and innovation firm based in Rio de Janeiro and Madrid. He was the first Brazilian to obtain an MSc degree from the University of Gastronomic Sciences in Italy. His current projects focus on native berries from the Atlantic rainforest, capsicum peppers, organic farming and olive oils. He is editorial adviser at PM Media, publishers of the magazine *Villas & Golf Gourmet*, in Lisbon.

Suzanne de la Barre is a postdoctoral research associate at the Department of Social and Economic Geography, Umeå University, Sweden. Suzanne is a long-time resident of the Yukon and living in the periphery has meant working as a cook in wilderness camps for tree planters and miners, running her own business in arts publishing and working on a variety of community development projects. Suzanne undertook her doctoral research in the Yukon where she examined the relationship between wilderness and cultural tourism guides, place identity and the goals of sustainable tourism. Her current research examines tourism and community development in mining intense circumpolar regions.

Julia Fallon is currently Head of MBA in the Cardiff School of Management, UWIC. She was previously Head of Centre for Tourism, Leisure and Events building on experience of teaching travel and tourism plus business subjects for over 20 years. Her research has been focused on historical leisure activity, particularly examining canals and waterways as leisure space using oral history as method. Living in Wales (where there is a heavy reliance on the spoken word) influenced her research approach as did speaking to people about their enthusiasm for canals and the inland waterways. These two factors now influence all her research.

Dawn Gibson is completing her PhD on the challenges of Indigenous community tourism operations in Fiji. Her Masters studied employee empowerment as a means of delivering consistent quality service within the cultural context of a local Fiji labour force. Her research interests include: sustainable tourism development, impacts of tourism, community/Indigenous tourism, backpacker and volunteer tourism, employee empowerment, service quality, hospitality and increasing local agricultural linkages to tourism through farm to table initiatives with hotels/resorts and local communities. She is currently a lecturer with the School of Tourism and Hospitality Management at the University of the South Pacific in Suva.

Esther Groenendaal is a lecturer in International Tourism Management Studies at NHTV University of Applied Sciences in Breda, the Netherlands.

Groenendaal studied Leisure Management (BBA) at NHTV and Environmental Sciences with a focus on Tourism and Leisure (MSc.) at Wageningen University. Her main research interests are entrepreneurial tourism concepts, migration issues in the EU, globalisation and social change. Groenendaal is coordinator of the bachelor programme 'Travel and Tourism Industry' and lectures in tourism impact studies, tourism and globalisation, and conceptual thinking. As a member of the NHTV Academy for Tourism, she lectures, organises workshops and develops curricula for tourism studies.

C. Michael Hall is a Professor in Marketing and Tourism at the Department of Management, University of Canterbury, New Zealand; Docent, Department of Geography, University of Oulu, Finland; and Visiting Professor, Linnaeus University School of Business and Economics, Kalmar, Sweden and the School of Hospitality and Tourism, Southern Cross University, Australia. He has wide-ranging interests in tourism, regional development, environmental history, environmental change and gastronomy.

Christopher Howard is originally from San Francisco, California, and is at present a PhD candidate in social anthropology at Massey University, Auckland. He holds a BA in Historical and Political Studies and Cultural Anthropology from Chaminade University of Honolulu, with an emphasis on East Asian religions and cultures. He also attended Sophia University in Tokyo, where he studied Japanese literature and contemporary society. Upon receiving his MA from Victoria University, Wellington, on language endangerment, narrative theory and critical discourse analysis, his interests shifted back towards anthropology and social theory – particularly issues surrounding globalisation and mobility. The topic of his doctoral thesis (in progress) is tourism and pilgrimage in the Himalayas.

Margo B. Lipman is a PhD candidate in the School of Business at James Cook University (JCU), Queensland Australia. Her main research focus is sustainable tourism and the ways in which tourism experiences impact participants' lives. She is also interested in the application of the Social Representations theory to the field of tourism. She has a Masters degree in Tourism from JCU as well as a Bachelor of Arts degree in Sociology from Princeton University.

Matthew McDonald is a lecturer at the University of Technology, Sydney and Chartered Psychologist with the British Psychological Society. He has held previous lecturing posts at Roehampton University, London and Assumption University, Bangkok. He is the author of three previous books *What To Do With Your Psychology Degree* (2008), *Epiphanies: An Existential Philosophical and Psychology Inquiry* (2009) and *Critical Social Psychology: An Introduction* (2011). His primary research interests include the application of continental philosophy to psychology, the social psychology of work and leisure, and career development and counselling.

Kevin Moore is a senior lecturer in psychology at Lincoln University, New Zealand. His research has focused on the psychology of leisure and recreation, tourist motivation, experience and decision-making, the psychology and social psychology of well-being and conceptual and theoretical issues in psychology and evolutionary psychology. He has published numerous book chapters, journal articles, conference papers and research reports in these areas over the past 20 years. His current work is focused on the links between leisure, tourism and well-being in the context of recent social trends and movements.

Apisalome Movono gained his undergraduate degree at the University of the South Pacific and is completing his Masters on the impact of tourism on communal development in Fiji, studying two resorts and their related villages. He has a keen interest in sustainability issues as they relate to local and Indigenous communities, together with the sociocultural and environmental impacts of tourism. He is currently a teaching assistant with the School of Tourism and Hospitality Management at the University of the South Pacific based in Suva, Fiji.

Meiko Murayama teaches at the University of Reading and the University of Gloucestershire and is also a visiting researcher at the Institute of Hospitality, Waseda University, Japan. Dr Murayama holds a PhD in tourism studies from the University of Surrey, UK and has worked at the University of Westminster and the University of Greenwich as a senior lecturer and as an associate professor at Nihon University, Japan. His main research interests have focused on tourism as a regeneration tool and more recent research interests include ethics in tourism, tourism planning, revitalisation of traditional destinations, comparative analysis and cultures and tourism.

Laurie Murphy is a senior lecturer in Tourism and Sports and Event Management and Business in the School of Business, James Cook University, Queensland, Australia. Laurie's main research focus is tourism marketing, with a focus on the backpacker market, tourist shopping villages, destination image and choice, and more recently destination branding. Laurie is on the editorial board of both the *Journal of Travel Research* and the *Journal of Travel and Tourism Marketing*.

Michael O'Regan completed a two-year research Masters of Business Studies at the University of Limerick, Ireland in 1997 and a PhD on backpacker mobilities in 2011 with the School of Service Management, University of Brighton, UK. Previously he has worked alongside the National Tourism

Development Authority of Ireland; Gulliver – Ireland's Information and Reservation Service; and as Marketing Executive for Wicklow County Tourism, Ireland. Michael has research interests in tourist, historic, sustainable and urban mobilities, particularly in relation to processes of globalisation and cosmopolitanism.

Fabio Parasecoli is Associate Professor of Food Studies at the New School in New York City. His research focuses on the intersections among food, media and politics. His current projects focus on the history of Italian food, food and masculinity in movies and on the socio-political aspects of geographical indications. He is programme advisor at Gustolab, a centre for food and culture in Rome, and collaborates with other institutions such as the University of Illinois Champaign-Urbana, Universitat Oberta de Catalunya in Barcelona and the University of Gastronomic Sciences in Pollenzo, Italy. Among his recent publications: *Food Culture in Italy* (2004), the introduction to *Culinary Cultures in Europe* (The Council of Europe, 2005) and *Bite Me! Food in Popular Culture* (2008). He is general editor with Peter Scholliers of a six-volume *Cultural History of Food* (forthcoming 2011).

Gavin Parker is Chair of Planning Studies at the University of Reading, UK. He is a chartered planner and has a research focus on countryside planning and policy, including rural economic development. He has published widely on a variety of topics under that umbrella, including rural tourism and leisure practices. He is the author of *Citizenships, Contingency and the Countryside* (Routledge, 2002) and *Key Concepts in Planning* (Sage, 2011). He maintains a specific interest in Japanese policy and practice and has been a visiting professor at the University of Tokyo on several occasions.

Stephen Pratt gained his PhD from the Christel deHaan Tourism and Travel Research Institute, University of Nottingham, UK. He is particularly interested in the economic impacts of tourism and destination marketing as well as issues of sustainability. He has published in a range of journals including *Annals of Tourism Research*, the *Journal of Travel Research* and *Tourism Analysis*. He is currently Senior Lecturer with the School of Tourism and Hospitality Management at the University of the South Pacific based in Suva, Fiji.

Sagar Singh has been associated with the Centre for Tourism Research and Development (CTRD) since 1984 and is author of two books on tourism, titled *Studies in Tourism: Key Issues for Effective Management* and *Shades of Green: Ecotourism for Sustainability*. He has written more than 21 research papers and book chapters and is author of the international bestselling books *Man: Essays in Anthropology* and *Hinduism: An Introductory Analysis*. He specialises in the anthropology of tourism and is currently Senior Research Associate at CTRD. His latest contribution is the article 'Ghost marriage: Magic among the Nuer', published in the journal *The Eastern Anthropologist*.

Marg Tiyce has been a researcher, tutor and associate lecturer in the School of Tourism and Hospitality Management at Southern Cross University since 1998. Her research and teaching focus has been in the areas of tourism theories and practices, special interest tourism, tourism planning and the environment, event management and tourism research. She is currently completing doctoral research on the experiences of long-term travellers in Australia. Her research probes the actions and deeper meanings of travellers' journeys as they relate to their wider lives.

Michael Wearing, Senior Lecturer in the School of Social Sciences and International Studies, Faculty of Arts and Social Sciences, University of New South Wales (UNSW), Sydney, Australia. He received a PhD in sociology from UNSW whilst a scholar at the Social Policy Research Centre and has gone on to teach and publish in the areas of social policy, sociology and political sociology while an academic at Sydney University and then UNSW. He is the author of several books and over 50 refereed publications. These include texts on sociocultural aspects of tourism, community services, social welfare and social policy. His current research interests are in the environment and ecotourism, the politics of welfare rhetoric, change in human service organisations and comparative social policy.

Stephen Wearing specialises in the social sciences in natural resource management and has degrees in environmental and town planning and his PhD focused on community development within the context of leisure and tourism. His research and publications range across the areas of the sociology of leisure and tourism. He has been project director for a range of natural resource management projects and a team leader for a variety of ecotourism, volunteer tourism and outdoor education activities internationally. His latest book *Tourism Cultures: Identity, Place and Traveller* contributes to the growing area of 'critical tourism studies'.

1 Starting Slow: Thinking Through Slow Mobilities and Experiences

Simone Fullagar, Erica Wilson and Kevin Markwell

Slow food, slow cities, slow living, slow money, slow media, slow parenting, slow scholarship and ... slow travel. It seems that wherever we look, the prefix 'slow' is being added to another sector, phenomenon or industry. Being slow was once an entirely derogatory term that signified one's inability to 'keep up' in the competitive spheres of work and leisure. Yet curiously, the meaning of slow is now starting to shift, as slowness today is invoked as a credible metaphor for stepping off the treadmill, seeking work–life balance or refusing the dominant logic of speed. Slowing down has become an antidote to the fast paced imperatives of global capitalism that urge the entrepreneurial self to speed up, become mobile and work harder in order to be valued as successful, productive and conspicuous consumers (Humphrey, 2010; Rose, 1999; Schor, 2010).

One only has to glance at recent television programming to note the increased interest in slow and alternative forms of travel. In addition to programmes detailing travellers trekking over vast landscapes, there has also been a proliferation of shows documenting intercontinental travels via vehicle. The most popular perhaps are *Long Way Down* (2007) and *Long Way Round* (2004), which follow celebrities Ewan McGregor and Charlie Boorman as they motorcycle over multiple continents engaging the locals at every opportunity.

Another indication of consolidation of this phenomenon is observed through the products and the services that are now available under the banner of slow tourism. Several websites make claim to the phenomenon offering 'slow travel' experiences, ranging from fully booked tours to long-stay accommodation. In addition, there is a range of full length slow travel guide books titled for different cities around the world, which state 'The Slow Guides are for anybody who wants to slow down and live it up. They celebrate all that's local, natural, traditional, sensory and most of all gratifying about living in each of these corners of the world' (Slow Guides, n.d.). Clearly, for those fed up with fast, the goals of slow are to explore the possibilities of being different, working differently, playing differently and, in the context of travel in particular, moving differently.

In this book, we ask: what do slow mobilities mean for tourism? What effects do slow mobilities have and how do they evoke different ways of engaging with people and place? And, how are we also 'moved' by slow travel experiences in ways that lead us to question, connect with and desire to know the world differently? This book arose from our shared interest in thinking 'through' the multiplicity of experiences and representations of slow travel. In both our personal and professional lives, each of us has strived – and continues to strive – to maintain a sense of slow, whether it be through a choice of rural and alternative lifestyles, installing solar panels, growing our own vegetable gardens, going part-time to look after young children, or trying to eat and travel in a more sustainable manner. We also yearn for a sense of slow scholarship, as we continue to question our roles and privileges in the knowledge production system that has become higher education. We wanted to embrace a critical ethos that questioned the unsustainable pace of consumerism, the demands of work and the desire for alternative mobilities (Fullagar, 2003; Humphrey, 2010; Sheller & Urry, 2006; Urry, 2002).

In this introductory chapter, we consider how 'slow mobilities' figure within the historical emergence of the slow living movement as a constellation of diverse ideas and cultural forms relating to food, cities, money, media and travel (Cresswell, 2010; Dickinson & Lumsdon, 2010; Honoré, 2005; Parkins & Craig, 2006; Tasch, 2008). With our focus on the experiences of travel and tourism we understand slow mobilities in Cresswell's sense as 'particular patterns of movement, representations of movement, and ways of practising movement that make sense together' (2010: 18). Slow ideas are permeating the contemporary tourism imaginary, eliciting a range of nostalgic and future oriented desires for local/global connectedness, low carbon options and journeys that value embodied experiences of time. A plethora of slow travel narratives, images and discourses now circulate globally through the popular press, travel blogs and magazines, as well as guidebooks, marketing for tours and destinations (see Funnell, 2010; Germann Molz, 2009; Sawday, 2009, 2010).

Slow tourism has been the focus of recent discussion in the tourism literature about how to conceptualise 'slow' in relation to the principles of sustainable tourism, as well as how to identify the range of slow practices, motivations and supply issues (infrastructure, regulation and markets) for tourism development (Dickinson & Lumsdon, 2010; Dickinson et al., 2010; Hall, 2009; Lumsdon & McGrath, 2010). Lumsdon and McGrath's research has identified some parameters around slow tourism in terms of 'slowness and the value of time; locality and activities at the destination; mode of transport and travel experiences; and environmental consciousness' (2010: 2). A number of typologies have emerged to categorise the environmental practices of slow tourists through metaphors of 'hard or soft' and 'heavy or light' (Dickinson & Lumsdon, 2010). Yet there is little consensus on what 'slow' actually means, and how it is practiced or interpreted in relation to different tourism contexts, cultures and mobilities.

Our aim in this book is not to attempt to pin down the mobile meaning of slow travel experiences, but rather to explore from different vantage points the dimensions of slow that draw out the complexities of local-global, time-space, nature-culture, self-other and personal-political relationships. Crucial to developing different insights the contributors to this book have also employed a range of research methods to explore questions of slow mobility and the mobility of meaning in tourism (Watts & Urry, 2008).

Experiencing Slow Mobilities

The notion of slow mobilities emphasises more than movement, or transport, between places. Rather, the term 'mobilities' encapsulates a range of spatio-temporal practices, immersive modes of travel and ethical relations that are premised on the desire to connect in particular ways and to disconnect in others. Slowness is more than anti-speed, however. Rather, slow is embodied in the qualities of rhythm, pace, tempo and velocity that are produced in the sensory and affective relationship between the traveller and the world (Cresswell, 2010). A slow relation to the world has been shaped by a number of social movements in specific parts of the world that have become mobile and virtual forms of social connection and identification.

In particular, the concept of slow travel has emerged from the Slow Food and Slow Cities (CittaSlow) movements that both originated in Italy in the 1980s and 1990s. The Slow Food movement was initiated in 1986 by Carlo Petrini as a response to the opening of a McDonald's restaurant in an area of cultural significance in Rome. As can be assumed, Slow Food was a collective retaliation against the phenomenon of increased global consumption of fast food. Officially constituted three years later in 1989, the movement has now expanded to 132 countries with 100,000 members of the Slow Food International organisation (Slow Food, 2010a). Slow Food International's mission statement is 'to defend biodiversity in our food supply, spread taste education and connect producers of excellent foods with co-producers through events and initiatives' (Slow Food, 2010b). As with Slow Food, Slow Cities focus on 'the development of places that enjoy a robust vitality based on good food, healthy environments, sustainable economies and the seasonality and traditional rhythms of community life' (Knox, 2005: 6). 'Slow Cities' has also become institutionalised as a movement (CittaSlow) and progressed into a topic of academic inquiry (Knox, 2005; Mayer & Knox,

2006; Parkins & Craig, 2006; Tasch, 2008). As Pink (2008: 97) explains, 'CittaSlow emphasises local distinctiveness in a context of globalisation and seeks to improve quality of life locally'.

Like the Slow Food and CittaSlow movements, slow forms of tourism embrace this emphasis on the local consumption of food that draws upon culinary heritage or organic principles, as well as the sensory embodiment of the journey (taste becomes as important as sight). Slow travellers are often distinguished by a desire to experience a different temporality to that of the 'bucket list' of places to stop over and move on from. Slow immersion in the particularity of place can evoke and incite different ways of being and moving, as well as different logics of desire that value travel experiences as forms of lived knowledge. Against the high environmental impact of the aeroplane and car, a range of s/low carbon modalities figure as alternatives (walking pilgrimages, canoeing, leisurely cycling, place-based experiences) that value nature and cultural traditions.

The multiple desires for slow travel often play out in complex and con-tradictory ways through discourses and narratives of travel. Slow travel is marketed in the representations of high status travel magazines (for example, the *Australian Gourmet Traveller*) to signify the accumulation of cultural capi-tal. The glamorous 'gourmet slow' traveller consumes high quality food and wine as a reflection of cultural taste achieved through commodified experi-ences that are distinguished from mass tourism. While commodified forms of slow constitute a niche market, it is not surprising that the slow move-ment has come under criticism as an elitist preoccupation of the harried middle classes (Heldke, 2003; Wilson, 2010).

Yet slow importantly signifies anti-consumerist displeasures associated with unsustainable lifestyles and eco-desires for different kinds of identities (Schor, 2010; Soper, 2008). In his book *Go Slow Italy*, Alastair Sawday (2009: 13) describes the shift towards slow politics as 'a bridge from panic to plea-sure'. In this sense slow mobilities can be understood as part of a broader 'life politics' (Rojek, 2010; Rose, 1999) where negotiations occur around values of freedom and responsibility, behaviours that are sustainable or con-sumerist, and social relationships that enable engagement rather than obser-vation, respect rather than exploitation and reflexivity rather than status seeking identities. Slow travel practices are informed by a diverse range of ethical sensibilities that bring together pleasurable modes of engaging with nature, or culture, and a politically reflexive sense of identity that is criti-cally aware of the impact of one's own tourist behaviour. As Dickinson and Lumsdon (2010) have identified in their research, however, there exist many tensions between an individual tourist's concerns about environmental impact and the desire for slow experiences in the context of a global tourism system premised upon economic growth. While slow travel practices reflect a range of ethical-political positions, they are yet to be fully explored in the academic literature despite the growth of industry and popular discourses.

The slow movement aims to revalue quality leisure time, sociality and non-consumerist experiences that aim to minimise the environmental footprint (Dawson *et al.*, 2008; Honoré, 2005; Mair *et al.*, 2008). Yet there is a tension that is not easily resolved within slow philosophies about the carbon footprint created by air or car travel (especially if one travels anywhere from Australia or New Zealand). Slow travel can contribute to debates about sustainable tourism and the search for alternative mobilities in relation to the pressing issues of peak oil, food security and transnational flows. However, it would be naïve to see slow as a simple answer to the broader issue of predicted growth in global travel and middle-class consumption in emerging economies such as India and China. The constellation of ideas that connect through the principles, philosophies and practices of slow mobility potentially offer creative and culturally diverse ways of moving in the world – both at home and away.

Structure of the Book

This book is organised in four major parts. Each chapter contributes conceptually, or empirically, to thinking through the multiplicity of slow tourism and travel (we use these terms interchangeably rather than perpetuate a dualistic conceptualisation). Contributors to the first part, 'Positioning Slow Tourism', consider questions about temporality and how time is experienced differently through slow mobilities. In their respective Chapters, 2 and 3, both Christopher Howard and Kevin Moore reflect critically on theories of time in relation to issues of well-being and pleasure that are evoked by travel desires for the good life and slow journeys such as secular pilgrimage. In Chapter 4 Stephen and Michael Wearing and Matthew McDonald contribute to a critical analysis of the commodification of the time-space of travel within global capitalism. They point towards the potential of ecotourism to generate sustainable and pleasurable experiences that connect tourists and host communities.

The second part of the book, 'Slow Food and Sustainable Tourism', is organised around the emergence of slow food tourism and its connections to sustainability and eco-gastronomy. In Chapter 5 C. Michael Hall provides a succinct summary of the Slow Food movement, critically examining its paradoxes and contradictions in relation to sustainable tourism. That is, how do we continue to emphasise the local, regional and environmental within the context of an ever-globalising and mobile world? As Hall warns us quite bluntly, we must ensure that 'slow' does not become merely another institutionalised method of 'screwing the Earth'. Turning to a more micro, nation-specific context, Fabio Parasecoli and Paulo de Abreu e Lima in Chapter 6 present a case study of how a group of local food producers, restaurateurs and media professionals in the Brazilian town of Paraty have launched a sustainable gastronomy programme. The Paraty chapter demonstrates how slow tourism can allow visitors to enjoy and participate in food production and culinary traditions 'as embedded and embodied performances of living cultures'. Moving to Australia, Margo Lipman and Laurie Murphy in Chapter 7 explore the growth of WWOOFers (Willing Workers on Organic Farms) in terms of the potential to connect sustainable food production with more environmentally-friendly ways of travelling.

'Slow Mobilities' is the third part of the book, consisting of four chapters which explore distinct examples of slow mobilities and ways in which tourists can experience the spatio-temporality of the journey. To begin with, in Chapter 8 Simone Fullagar offers a gendered approach to slow travel through her ethnography of Australian women's experiences of an annual mass cycle tour event as a form of 'alternative hedonism'. Cycling as a slow mobility is defined here as much more than a means of transport from A to B; rather, it offers a potentially transformative experience or journey of self revelation about the social and natural world. In Chapter 9 Marg Tiyce and Erica Wilson also employ ethnography, this time to document the experiences of long-term travellers who define themselves as 'wanderers'. These wanderers drive around the country, in search of a slower pace, time and speed that might offer up a sense of meaning, well-being or way of engaging more deeply with people and place. This wandering also allows travellers a sense of resistance to the 'status quo' of fast-paced life back at home.

In Chapter 10, Michael O'Regan, explores the tourist habitus of hitch-hiking as a self-powered mobility through the tension that exists between slow and competitive desires (European hitch-hiking competitions) that embraces risk taking, local engagement and mastery. Julia Fallon in Chapter 11 completes the focus on slow mobilities in this section of the book by documenting the history of canal development and canal tourism in England. Fallon argues that the very nature of moving by canal, particularly given their narrow structure in much of the country, deliberately encourages a slow mobility, where the traveller can become more relaxed and in tune with the environment around them.

The final part of the book, 'Slow Tourism Places', is organised around the theme of tourist destinations and places that represent some of the concerns identified by the Slow Cities movement. Taking readers to a very particular climate and culture in Chapter 12, Suzanne de la Barre writes about the marketing of 'Yukon time' as a destination value in northern Canada. Questions about the process of othering Indigenous peoples are raised in this analysis of how slow travel values are used in marketing discourses to reformulate potentially negative infrastructural deficiencies or cultural idiosyncrasies as quaint and charming aspects of life or travel in the territory. In Chapter 13, the theme of bridging traditional and contemporary cultures is also addressed by Meiko Murayama and Gavin Parker in their analysis of fast and slow Japan. Tourism authorities have begun to realise the potential of slow travel in marketing rural tourism as a part of a regeneration strategy.

In Chapter 14, Dawn Gibson, Stephen Pratt and Apisalome Movono explore how sustainable practices are experienced by tourists taking part in the 'Tribewanted' project on the island of Vorovoro in Fiji. Learning about traditional Indigenous customs, food production and consumption practices, tribe members express a deeply felt and memorable connection to place and people through 'slow' community-building activities. Slow travel experiences are also recreated through a virtual tourist community online where the immediacy of time and space are transcended. Exploring a cross-cultural context in Chapter 15, Esther Groenendaal examines why Dutch tourism lifestyle entrepreneurs have moved to France to open B&B accommodation that reflects a slower pace of life. She considers how these personal choices about working in tourism are also shaped by broader socio-political movements that value culture, creativity and environment. As the final contribution to the book in Chapter 16, Sagar Singh explores the significance of traditional and more modern forms of slow tourism (pilgrimage, yoga tourism) within Indian culture and history. Western misconceptions of Eastern practices, such as yoga, are also explored in ways that question ethnocentric assumptions about what slow means and how it figures in a globalised tourism market.

In conclusion, the collection of chapters in this book broaden and deepen the academic research on slow tourism, demonstrating the connections, contradictions and complexities inherent in the concept of 'slow' as it relates to travel. Drawing on a range of disciplines including sociology, anthropology, history, food studies, cultural geography and cultural studies and tourism/hospitality management, the contributors to this book also reveal the diverse and multidisciplinary nature of slow travel. We hope that this collection will add to the body of knowledge concerning this emerging tourism phenomenon which, we believe, has the potential to challenge the ways that tourism is performed and organised.

References

Cresswell, T. (2010) Towards a politics of mobility. *Environment and Planning D: Society and Space* 28 (1), 17–31.

Dawson, D., Karlis, G. and Heintzman, P. (2008) Slow living: Postmodern temporality, the European experience, and the Sabbath. 12th Canadian Congress on Leisure Research, Montreal, Concordia University, 13–16 May, 2008.

Dickinson, J. and Lumsdon, D. (2010) *Slow Travel and Tourism*. London: Earthscan.

Dickinson, J., Lumsdon, L. and Robbins, D. (2010) Slow travel: Issues for tourism and climate change. *Journal of Sustainable Tourism* 19 (3), 281–300.

Fullagar, S. (2003) On restlessness and patience: Reading desire within Bruce Chatwin's narratives of travel. *Tourist Studies* 4 (1), 5–20.

Funnell, A. (2010) The slow movement. *ABC Radio National, Future Tense,* 2nd September 2010.

Germann Molz, J. (2009) Representing pace in tourism mobilities: Staycations, slow travel and The Amazing Race. *Journal of Tourism and Cultural Change* 7 (4), 270–286.

Hall, C.M. (2009) Degrowing tourism: Décroissance, sustainable consumption and steady-state tourism. *Anatolia: An International Journal of Tourism and Hospitality Research* 20 (1), 46–61.

Heldke, L.M. (2003) *Exotic Appetites: Ruminations of a Food Adventurer.* New York, London: Routledge.

Honoré, C. (2005) *In Praise of Slow: How a Worldwide Movement is Challenging the Cult of Speed.* London: Orion.

Humphrey, K. (2010) *Excess: Anti-consumerism in the West.* Cambridge: Polity.

Knox, P. (2005) Creating ordinary places: Slow cities in a fast world. *Journal of Urban Design* 10 (1), 1–10.

Long Way Round (2004) D. Alexanian & R. Malkin (producers). Elixir Films and BBC TV.

Long Way Down (2007) D. Alexanian & R. Malkin (producers). Elixir Films and BBC TV.

Lumsdon, D. and Mcgrath, P. (2010) Developing a conceptual framework for slow travel: A grounded theory approach. *Journal of Sustainable Tourism* 19 (3), 265–279.

Mair, H., Sumner, J. and Rotteau, L. (2008) The politics of eating: Food practices as critically reflexive leisure. *Leisure/Loisir* 32 (2), 379–405.

Mayer, H. and Knox, P.L. (2006) Slow cities: Sustainable places in a fast world. *Journal of Urban Affairs* 28 (4), 321–334.

Parkins, W. and Craig, G. (2006) *Slow Living.* Sydney: UNSW Press.

Pink, S. (2008) Sense and sustainability: The case of the Slow City movement. *Local Environment* 13 (2), 95–106.

Rojek, C. (2010) *The Labour of Leisure.* London: Sage.

Rose, N. (1999) *The Powers of Freedom: Reframing Political Thought.* Cambridge: Cambridge University Press.

Sawday, A. (2009) *Go Slow Italy.* Bristol: Alastair Sawday Publishing.

Sawday, A. (2010) *Go Slow France.* Bristol: Alastair Sawday Publishing.

Schor, J. (2010) *Plenitude: The New Economics of True Wealth.* New York: The Penguin Press.

Sheller, M. and Urry, J. (2006) The new mobilities paradigm. *Environment and Planning A* 38, 207–226.

Soper, K. (2008) Alternative hedonism, cultural theory and the role of aesthetic revisioning. *Cultural Studies* 22 (5), 567–587.

Slow Food (2010a) Home page, accessed 23 April 2010. http://www.slowfood.com/

Slow Food (2010b) Our mission, accessed 23 April 2010. http://www.slowfood.com/

Slow Guides (n.d.) Accessed 21 April 2010. http://www.slowguides.com.au/

Tasch, W. (2008) *Inquiries into the Nature of Slow Money: Investing as if Food, Farms and Fertility Mattered.* White River Jct, Vermont: Chelsea Green.

Urry, J. (2002) Mobility and proximity. *Sociology* 36 (2), 255–274.

Watts, L. and Urry, J. (2008) Moving methods, travelling times. *Environment and Planning D: Society and Space* 26, 860–874.

Wilson, E. (2010) Beyond beans and cheese: Representations of food, travel and Mexico City in the *Australian Gourmet Traveller. TEXT* 14 (2), http://www.textjournal.com.au/

Part 1
Positioning Slow Tourism

2   Speeding Up and Slowing Down: Pilgrimage and Slow Travel Through Time

Christopher Howard

We are pilgrims through time
Augustine
(in Bauman, 1996: 20)

The path of modernity is one characterised by acceleration, interconnection and mobility. Amidst the increasingly fast tempo of life in late modernity, contemporary pilgrims and other 'slow travellers' express needs and desires for alternative experiences of temporality, while subverting the dominant 'cult of speed'. My objective in this chapter is to focus on issues of temporality in exploring three aspects of slow travel: the potential contradictions it presents, its direct relationship to modernisation and its link to pilgrimage. First, I locate slow travel in the present, teasing out certain tensions and paradoxes that are bound up with its situatedness in the socioculturally ambiguous conditions of late modernity (Bauman, 2000, 2008; Beck, 2006; Giddens, 1990, 1991; Heaphy, 2007). Second, as travel always occurs within greater socio-historical contexts, I suggest that contemporary slow travel and pilgrimage are contingent upon the radical transformations occurring in Western modernity (Bauman, 1996: 20; Castells, 2004; Hassan, 2008, 2009; Swatos, 2006; Urry, 2002, 2007). Third, although slow travel appears as a newly emerging trend, I discuss how it can be viewed in relation to one of the oldest – and most enduring – forms of human mobility: pilgrimage.

Cultural Contradictions of Slow Travel

Victor and Edith Turner (1978: 38) assert that modern pilgrimages may be read as 'meta-social commentaries' on the troubles of the epoch and a search for vanishing virtues. With this perspective in mind, I suggest that the emerging trend in slow travel may also represent such a commentary. However, in the late modern context, certain paradoxes present themselves. Connected to what Daniel Bell (1978) describes as the 'cultural contradictions of capitalism', pilgrims and 'slow travellers' from the 'fast world' (p. 1) (i.e. highly developed countries) rely heavily on modern technology and the global travel system to research, plan and transport themselves around the globe. Highly mobile subjects – largely from more affluent countries and backgrounds – today travel great distances, often to take largely inward journeys: to practice 'simplicity' and 'slowness' and experience 'authenticity'. They use the tools of the information age and the modern luxuries of travel to temporarily escape the voracious pace of late modernity, or what Virilio (2010: 88) refers to as the 'dromosphere' (dromos from the Latin term for 'race'). Displaying what Luhmann (1982: 305) calls 'temporal reflexivity', slow travellers and pilgrims choose to momentarily step out of the so-called 'dromosphere', though their points of departure will inevitably be their points of return. In an age of speed and information, if a computer engineer in Silicon Valley desires to make a traditional Buddhist pilgrimage in the Himalayas, or a couple from England desires to ride the Trans-Siberian railway – for example – such 'slow travel' can all be arranged with a few clicks of a mouse. For many pilgrims and other 'slow travellers' who live primarily in the 'fast world', such travel may represent one of the few outlets for temporary deceleration and distance from the pace and complexity of late modern life. If slow travel is to be read as a meta-social commentary, it is indeed necessary to consider the socio-historical factors which cause its emergence and inform its practice.

Fundamental to such an endeavour is the recognition that perceptions and practices of slowness are contingent upon those of speed and vice versa. While the slow movement appears 'new' in certain respects, it is best seen from a historical perspective as a continuation of the subversion of speed and the advocacy of simplicity that begins in the West with the Romantics and Transcendentalists. Moreover, slow ideology appears to be connected to the 'valued ideals' that define a pilgrimage. Crossing līmens (literally, 'thresholds'), pilgrims and other 'slow travellers' temporarily liberate themselves from the immediacies of 'fast' life – from email, cell phones and overflowing daily planners. Such liberation presents the possibility for alternative experiences of time, place and self – giving forth new modes of being-in-the-world. Pilgrimage and slow travel – with their emphasis on mindfulness and revitalising the relation between self and world (and possibly God(s), in the case of religious pilgrimage) – are thus best understood as a form of mobile-existential praxis.[1] As Castoriadis (1998: 77) reminds us, 'the very object of praxis is the new'; in the case of pilgrimage and slow travel, being anew. Moreover, this mobile-existential praxis carries meanings and implications on both individual and broader sociocultural levels. On one hand, a pilgrimage or other slow journey may signify a very personal endeavour aimed at introspection and self-transformation, while on the other, may be seen as a meta-critique and an indirect subversion of the existing social order.

While the 'new' is always contingent on what comes before, it is clear that both in theory and practice, issues of temporality are at the core of pilgrimage and slow travel, as well as the more general slow movement. Parkins (2004: 368) observes that 'slow living at its best envisages more than just a redistribution of time and an increase in leisure', but points towards alternative understandings of time itself. The founder of the slow food movement, Carlo Petrini (2001), explains that 'slow' means being in control of the rhythms and tempo of one's own life. Author of the international bestseller, *In Praise of Slowness*, Carl Honoré (2005: 44), describes the slow movement as 'a cultural revolution against the notion that faster is always better' and an attempt to 'fix our neurotic relationship with time'. He adds, however, that 'it is not about doing everything at a snail's pace' but 'seeking to do everything at the right speed' (Honoré, 2005). Although a universal 'right speed' is a problematic assertion, advocates of 'slow' raise salient questions about contemporary perceptions and understandings of time. But while this 'new politics of time' appears as a recent phenomenon connected to the zeitgeist of late modernity, the slow movement has its roots in various countercultural movements, beginning first with the European Romantics and American Transcendentalists in the late 18th and 19th centuries. Furthermore, aspects of the dissident ideology of 'slow' can be found in the Dadaist and Beat movements, as well as the Hippy and New Age movements from the 1960s onwards. Common to these movements is the general questioning and subversion of cultural hegemony, which since the industrial revolution has also meant questioning speed and dominant notions of 'progress'. Questions about the nature of time and debates on how it can best be used have of course existed for much longer, however, with myriad interpretations offered from cultures and civilisations around the world.

Temporality: Singular or Plural?

Although time in late modernity is almost incontrovertibly accepted as simply what the clock reads, looking to the ancient Greeks and subsequent philosophers, along with various Indigenous cultures around the world, we can see a vast multiplicity of understandings (Adam, 1994, 2004; Fabian, 2002; Melucci, 1998; van Loon, 1996). The Greeks, for instance, distinguished between two distinct types of time – *chronos* and *kairos*. In *Physics*, Aristotle and Jowett (2004) describes *chronos* as precise, quantifiable time, as in the passing of time in successive readings of a clock, or what Heidegger (Heidegger *et al.*, 2006: 458) refers to as the uniform and homogenous 'time of everyday'. *Kairos*, on the other hand, represents a time outside of such time, or moments when the ordinary flow of *chronos* is ruptured and things come into focus in a unique way. Mythical 'sacred time' is similarly depicted by Eliade (1954, 1957) as a timeless, eternally reoccurring present (*illo tempore*) in which all divisions between space, time and subject ultimately disintegrate. *Kairotic* or 'sacred' time, at least according to the philosophers, theologians and mythology, is nothing less than the moments that ultimately give meaning and value to life. Crossing the boundaries, or *līmens*, separating *chronos* from *kairos*, or from the profane to the sacred, is largely a matter of transcending time.

The ancient Greeks, for this reason, saw leisure time as the basis of culture and the highest good. In *Politics*, Aristotle (Aristotle & Jowett, 2004: 14) contends that leisure (*skhole*) is the 'first principle of all action'. Leisure, or *skhole*, is defined by Bourdieu (2000: 1) as 'the free time, freed from the urgencies of the world, that allows a free and liberated relation to those urgencies and to the world'. For the Greeks, *skhole* was not understood as mere idleness or wasting time on frivolous activities, but as a moderated condition in which self-perfection was to be sought in philosophy, athletics, music and the arts (Holba, 2010; Pieper & Malsbary, 1998). Leisure – a contested concept pertaining to how time is lived and experienced and whose Latin etymology literally means 'to be permitted' or 'to be free' – is inexorably bound up with notions of the good life and human freedom as debated upon in the West at least since the Greeks. While the filling of leisure time in late modernity encompasses a great range of activities, travel (as the exercise in freedom of movement in both time and place/space), is arguably one of the most significant contemporary symbols of leisure, which in turn symbolises freedom and the good life (Blackshaw, 2010). Expanding upon this notion, slow travel appears to advocate that it is not enough to merely gaze upon – and hence consume – sites, peoples and places. In the heritage of Epicurean philosophy, rather, one must experience travel, like food, art or friendship, deeply and sincerely with one's *whole being* for it to be meaningful and authentic. The key ingredient for such 'meaningful' experiences, then, is, first and foremost, time.

The Acceleration of Modern Life: From *Chronosphere* to *Dromosphere*

Although leisure time is indeed part of modern societies, it has typically been associated with routine breaks from paid employment. Under modern conditions, *chronos* time comes to dominate, while speed and acceleration become 'the defining experience of modernization' (Parkins, 2004: 365). Classical social theorists such as Weber, Marx and Simmel observed the 'time is money' attitude pervading the modern era, an attitude characterised by an obsession with speed, efficiency and rational calculation. Weber (2003: 181) sees this highly regimented, essentially militaristic style of life consigning individuals to an 'iron cage', marked by spiritual disenchantment and a fundamental loss of meaning and purpose in life. Marx *et al.* (1977) examine the commodification of time and discuss how mechanised modes of production and wage labour lead to human alienation. In 'The Metropolis and Mental Life' (Simmel et al., 1997), Simmel observes how the excessive punctuality, calculability and exactness resulting from the money economy, in conjunction with the overstimulation of nerves in the modern metropolis leads to what he calls a 'blasé attitude'. The essence of the 'blasé attitude', he explains, is 'an indifference toward the distinctions between things', though not in terms of actual perception, but of value and meaning. Pheno-existential philosopher and author of the monumental work, *Being and Time* (*Sein und Zeit*, 1927), Martin Heidegger (Heidegger et al., 2006) develops the most extensive critique of modern clock-time, however, claiming that this form of artificial temporality distorts a larger 'reality' and prevents us from realising 'authentic being' (*Eigenlich*). This failing, according to Heidegger, leaves us in a state of existential limbo and alienation (*Enfrembudung*). Amidst the endless distractions encountered in modern *chronos* present, Heidegger asserts that subjects forget the fundamental temporal constitution of their existence. 'The consequences of this forgetfulness', Guignon and Aho (2010: 35) observe, 'is a uniquely modern way of being, characterized by what Heidegger calls "acceleration" (*Beschleunigung*), a kind of "mania" in which we try to cram in as many novel sensations and experiences into a particular span of time as we can, resulting in an inability to be quiet and still'. With quiet amusement I find myself wondering what Heidegger would make of the hyperconnected, media-saturated age of the iPhone, Facebook and Twitter we are witnessing at present. Despite the omnipresence of the internet and the trend in ceaseless digital communication, as with all major social trends, counter currents also exist; the popularity of pilgrimage and slow travel – with their emphasis on simplicity, mindfulness and embodied experience – being a case in point.

A century ago, Simmel (Simmel et al., 1997: 171) noted the peculiar human tendency towards interconnectedness, or what he called the 'will to connection'. The problem with the open-ended and increasing speed of networked communication is that it leads to what Hassan (2008: 184) calls 'abbreviated thinking', defined as 'a form of dealing with information that is necessarily surface-level because of the sheer volume of information we are confronted with and the time constraints which social acceleration places upon us'. Paul Virilio (1997) claims that the speed and information that epitomise late modernity lead people to experience a 'fundamental loss of orientation in the world'. Moreover, he claims that what is effectively being globalised today is not so much space, as time, and that for the first time, history is unfolding within a one-time-system, which he calls 'global time' (Virilio, 2006b: 91). Under a 'dictatorship of speed', run by neo-liberal governments, mass media and transnational corporations, 'global time' allegedly prefigures 'new forms of tyranny' and results in the 'time poverty' that characterises much of the developed world. Plausibly, it is the desire to exit such a reality and take time into one's own hands, even if temporarily, that informs slow travel and many contemporary pilgrimages.

Critiques of modern acceleration and the hegemony of *chronos* time put forth by Weber, Simmel, Heidegger, Virilio and others, is that such a mode of existence inhibits subjects' capacities for reflexive thought, judgement and moral responsibility and ultimately hinders human teleological potential. It is not only the philosophers and critical social scientists who question the hegemony of speed, however, as demonstrated by the emergence of grass roots organisations such as Slow Food International, the Society for the Declaration of Time in Australia, the Sloth Club of Japan, and 'Take back your time day' in the United States. It is important to keep in mind, however, that 'speed creates slowness' (Parkins, 2004). In other words, 'the experience and value of slowness' is historically contingent upon modern notions of speed. During the industrial revolution, as new technologies increased speeds of production, along with the accompanying consumption patterns and the overall pace of life – 'simplicity' and 'slowness' began to be idealised as part of a lost, utopic past by the Romantics in Europe and the Transcendentalists in America.

In the late 18th and 19th centuries, for writers such as Rousseau, Wordsworth and Goethe in Europe, and Thoreau, Emerson and Whitman in America, leisurely wanderings in the open air – away from the corruption and decadence of the city – became a 'valued ideal', as well as a moral and political statement (Urry, 2007: 78). By the late 18th century, 'the appreciation of nature, and particularly wild nature, had been converted into a sort of religious act' (Mathieu, 2009: 347). Rousseau *et al.* (1997), often acknowledged as the first romantic, is largely credited with the modern sacralisation of nature. His European-wide best-selling novel *La Nouvelle Heloise* of 1761 sent the educated classes on literary-nature pilgrimages to Lake Geneva, where mountain landscapes were venerated for their (only recently acknowledged) 'sublime' beauty. Later in life, in *The Reveries of a Solitary Walker* (Rousseau & Butterworth, 1992), Rousseau recounts his greatest pleasure as laying back in a rowboat and allowing it to drift freely under an endless sky. In his essay, 'On Walking' Thoreau (2007: 12) similarly expressed the valued ideal of rural solitude and unhurried movement in nature:

> I think that I cannot preserve my health and spirits unless I spend four hours a day at least – and it is commonly more than that – sauntering through the woods and over the hills and fields absolutely free from all worldly engagements.

While solitary walks in the country were previously considered dangerous and rebellious acts, by the late 19th and early 20th centuries, walking, cycling and hiking in the 'open air' became increasingly popular with the bourgeoisie (Urry, 2007: 79–81). Although Romantic and Transcendentalist authors appeared to have captured the emotional tensions of the rapidly transforming times, somewhat paradoxically we can see that modern technology and speed also helped facilitate the growing enthusiasm for 'wild nature' by way of making it more accessible to increasingly mobile city dwellers. In his essay, 'The Alpine Journey', Simmel (Simmel et al., 1997: 219) discusses how the completion of railways in Switzerland led to 'the whole-sale opening-up and enjoyment of nature', whereas remote destinations could have previously only been reached on foot. With the ease of travel, Simmel observes how the Faustian wish to scale a mountain peak and pro-claim 'I stand before you, nature, a solitary individual' is thwarted by capi-talistic enterprise and a nascent mass culture. One could say that the 'wholesale opening-up' of the Swiss Alps made possible by technological innovations, as well as increases in the leisure time and excess capital of the middle classes, is an early example of the democratisation of tourism that would truly come to fruition post World War II (MacCannell, 1976).

In modernity, railway travel and the colonisation of space and time indeed became a symbol of progress, but also of potential destruction and de-humanisation (Parkins, 2004: 365). John Ruskin et al. (1956: 159) famously lamented how travel by rail made him feel like an inanimate parcel, and advocated that the only way to appreciate natural landscapes was on foot. Urry (2000: 54) notes that as the railways and other modern modes of transportation arose, people became increasingly able to compare and contrast different forms of mobility. Thus, as walking and horseback become romanticised and aestheticised in modernity, fast forward to the age of air travel and the 'slow' mobility of travel by rail and ship. Eriksen (2001: 54) notes that in recent times, rail travel is often celebrated as 'a contemplative, quiet alternative to the hectic bustling of air travelling and the frustration of driving'. It thus appears that in late modernity, earlier forms of mobility – including the earliest of all – bipedalism – are seeing renewed cultural significance.

Pilgrimage and the Quest for Slowness

Pilgrimage, though an ancient practice, in recent times has become an increasingly popular avenue for subjects of the 'fast world' to both tempo-rarily escape and critique the 'cult of speed'. While pilgrimage has played a central role in the history of human mobility, it also represents a paradig-matic form of slow travel. Despite strong associations with organised reli-gions, pilgrimage is by no means limited purely to the realm of institutionally sanctioned religious practice. In fact, in recent times adapted forms of quasi-religious or secular pilgrimages are flourishing in many parts of the world (Arellano, 2007; Badone & Roseman, 2004; Collins-Kreiner, 2010; Timothy & Olsen, 2006; Reader, 2007). Characteristic of more recent approaches are the 'postmodern' blending of religious traditions and sites with forms of New Age spirituality, the sacralisation of certain cultural or natural sites, as well as the incorporation of the global tourism industry and information technologies (Bittarello, 2006; Rountree, 2005). Although such approaches may stretch the boundaries of traditional conceptions of pilgrimage, they nonetheless follow the perennial 'pilgrimage model'. As Behera (1995: 44) points out, 'every culture has had its archetypal quest, and in every age, this search has been given expression in journeys to places that embody the higher values of the culture. Man seeks the tangible symbols of his ideals'.

Building on this universalistic conception, I employ Morinis' (1992: 4) definition of pilgrimage as 'a journey undertaken by a person in quest of a place or a state that he or she believes to embody a valued ideal'. This notion of pilgrimage encapsulates diverse forms of journeys which need not be limited to institutionalised religion – indeed to religion at all – or even to specific physical destinations. In her research on the connections between early Christian pilgrimage and modern tourism, Adler (2002: 30) makes the important point that 'even innovative travel institutions and styles rest upon some pre-existing conventions'. Thus, while the 'pilgrimage model' persists, over time socio-historical developments gave way to updated forms of questing, such as the European 'grand tour', the exploration and conquest of uncharted, exotic lands, romantic flights from 'civilisation' and eventually certain contemporary modes of travel (e.g. eco, backpacking, voluntourism, spiritual and slow tourism) (Clifford, 1997; Cohen, 2004; Graburn, 1983).

Pilgrimage as a quest towards one's 'valued ideals' may in late modernity be made to museums or historical monuments, ancestral homelands (real or imagined), cultural events or to the homes of revered authors or celebrities (Campo, 1998; Dubisch, 2004). In late modernity, world heritage sites and so-called 'power places', marked by perceived natural beauty, cultural significance and 'spiritual magnetism', such as Machu Picchu, Delphi or the Himalayas have also become global pilgrimage centres (Arellano, 2007; Preston, 1992; Singh, 2005). While the forms of pilgrimage may change under late modern conditions, the meaning – as 'the typically human desire to seek out the sacred' – stays largely the same, though as Tomasi (2002: 20) points out, 'what symbolizes or articulates "the sacred" today, may be different from the past, even at the same site ...'.

While a specific destination or 'sacred centre' is traditionally thought to be the ultimate goal of a pilgrimage, for many pilgrims what facilitates inner transformations is the act of journeying itself, with arrival being potentially anticlimactic (Frey, 1998). Turner (1973) discusses how being 'on the road' puts pilgrims (and tourists) in positions of anti-structure and liminality. The Latin *līmen* of liminality refers to a threshold or passageway, a state of being between two different existential planes and is similar to the Sanskrit concept of *tirtha* (part of the term for pilgrimage), literally meaning to cross a river ford (Bhardwaj, 1973). Liminality refers to being situated in an ambiguous, often ahistorical condition that exists outside of profane time and social structures and is thus characterised by a sense of being 'neither here nor there', but 'betwixt and between' normal social positions (Turner, 2007: 95). Such a precarious position makes subjects vulnerable and 'in a sense, dead to the world', while also liberating them from structural obligations (Turner, 1992: 49), in turn opening up new possibilities for *being-in-the-world*. Graburn (1983: 11) similarly observes how tourism, much like pilgrimage and rites of passage, involves 'separation from normal "instrumental" life and the business of making a living' and offers entry into alternative states 'in which mental, expressive and cultural needs come to the fore'. A slower tempo, whether it simply allows for temporary escape and reprieve from the pressures of fast life or allows for experiences which facilitate deep personal reflection, renewal and growth (or both), appears to be such a need arising in late modernity.

In her in-depth research on the Santiago de Compostela pilgrimage, Frey (1998: 72) explains how when pilgrims begin to walk the 'camino', their perceptions of the world change, including their sense of time, awareness of their bodies and the subtleties of the landscape. Slavin (2003: 13), who also conducted research on the Santiago pilgrimage, observes that 'distance from the everyday was created through an active avoidance or rejection of certain things associated with modern life such as commercialism, industrial landscapes, highways and noise'. Interestingly, many of Slavin's informants described themselves as being on 'spiritual journeys', though most expressed apprehension about their final arrival at the shrine in Santiago, which they associated with tourism, commodification and organised religion. For many pilgrims, the essence and meaning of the experience appears to be *in* the movement, along with being removed from the noise, speed and pressures of life at home. Elsewhere, Maoz (2004: 114), reporting on young Israeli backpackers in India, found that in taking breaks from stressful careers or after just exiting compulsory military service, many respondents claimed not to be 'doing anything' in India. Instead, they explained that in the 'land of spirituality', they were simply 'concentrating on being'. By engaging in the practices of yoga and meditation, as well as attempting to immerse themselves in the local culture – perceived as spiritually superior and more authentic – many Israeli travellers expressed desires to 'find themselves' during their journeys. If we employ Morinis' (1992) definition, then slow journeys such as these become pilgrimages – the sought after 'valued ideal' being authentic selfhood.

In my own research on Himalayan travel, I have found similar themes. One interviewee described his four-week trek to the base of Mount Everest very much in the parlance of a pilgrimage and claimed that it was the best that he had done in his life. When asked what made it so, he emphasised the combination of awe-inspiring scenery, the physical challenge of trekking in high altitude, the way the Sherpa culture lives in harmony in this dramatic environment and the simplicity of life as one treks towards a meaningful goal. As he explains:

> Life simplifies right down to waking up, walking from A to B, eating, and sleeping. This gives you time away from the everyday stresses and allows you to think about the more long term aspects of life. That's not to say I don't think about these things at home, but as you don't have all the other problems of life crowding your head. You have more 'mental' space to think. Although, I think a lot of the time I probably wasn't thinking about much at all. There is no internet (well apart from Namche Bazaar), no cell phone coverage. You feel you can get away from all the social pressures in life (all the people you should be emailing or contacting), as well as all the work commitments. And you have a good excuse – there's no coverage so you cannot contact me!

While being in a liminal position, temporarily removed from normal social structures and commitments, played a part in the enjoyment of his experience, having a concrete goal also held significance:

> Although life is simplified and at a much slower pace, I don't recall ever being bored. Because you have a definitive goal that drives you forward. I also had books to read. Views to look at, photos to look through, fellow walkers to talk to and play cards with. Just sitting in the sun, looking at a view, with a cup of tea or beer in hand with not a care in the world. And you also feel you're on a mission to accomplish something worthwhile and life-changing. It's not just simplifying life and slowing down. You could do that by being unemployed and being on the benefit. But then you would have no goal, no sense of achieving anything, and these are things I think humans need to feel as if they are flourishing and to make them happy.

One respondent from Tokyo – who has made multiple, extended trips to the Himalayas – explained how after finishing university and working for a few years he apparently came to a crossroads:

> Everything was getting fucked up ... my work was going nowhere, my band wasn't progressing and my drinking was going up and up. Many of my friends had gone travelling to India and I envied them. So, I quit my job and went to India for six months. I wanted to do a real hardcore backpacking trip.

On this trip and subsequent ones, he travelled to Buddhist pilgrimage centres[2] (though he did not identify himself as a pilgrim, noting that he was 'too rational') and spent a large portion of time in Dharamsala – the Dali Lama's home in exile – studying Tibetan Buddhism. In terms of the motives behind his travels, he expressed a yearning to believe that 'sacred places' and an 'ideal village' exist and the closest things could be found in the Himalayas. When asked what constituted an 'ideal village', he explained that it is a place where people live in harmony with the natural rhythms of life, are deeply spiritual and non-materialistic; as he put it, 'the opposite of Tokyo'. Like other similar traveller discourse, there was great emphasis put on simplifying life through a slowed temporality while simultaneously countering the desire to accumulate commodities – all of which again points towards certain ideals of purity, freedom and the quest for authentic selfhood.

Concluding Remarks

These accounts by two Himalayan travellers reflect the themes I have touched upon in my brief discussion of slow travel and pilgrimage as a meta-social commentary on the speed and complexity of late modern life. Slow travellers and pilgrims from the 'fast world' express needs and desires for slower tempos and simplicity in order to experience the self and the world at deeper and allegedly more authentic levels. At the same time, as I argued that the meaning and value of 'slow' is contingent upon the speed that marks modernity, it appears that in late modernity – characterised by unprecedented acceleration, interconnection and mobility – the 'valued ideals' that define a pilgrimage for many become an awakened state of *being-in-the-world* and the experience of authentic selfhood. In order to reach 'authentic' destinations where such experiences can be had, however, many slow travellers are more than happy to use the fast tools of the information age and the luxury of air travel to transport themselves there (as almost any slow travel blog demonstrates). Certain contradictions and ambiguities thus present themselves when temporarily leisured subjects from the 'fast world' endeavour to free themselves from the 'urgencies of the world'. With often limited time frames for slow travel due to mounting numbers of demands – real or self-perpetuated – in late modern life, subjects aim to maximise their slow experiences. In other words, in displays of 'temporal reflexivity', slow subjects often speed up in order to slow down.

Taking time into their own hands, pilgrims and slow travellers perform a mobile-existential praxis in which new possibilities for experiencing the relations between the self and world unfold. Under the accelerated social conditions of late modernity, such praxis may crucially allow for *kairos* – understood as the moments which add value and meaning to life – to emerge. The next question for future research on pilgrimage and slow mobilities then, is what happens when slow subjects return to the fast world? In other words, how and to what extent will this meta-social commentary and mobile-existential praxis effect change – both of self and society? As always, only *time* will tell.

Notes

(1) Here I understand praxis as reflection and action taken in order to transform oneself and the world. As opposed to what Hannah Arendt (1977: 6–7) calls the *vita contemplativa*, mobile-existential praxis refers to the *vita activa*.

(2) However, he did not identify himself as a pilgrim, noting how he felt he could never fully integrate with the 'spiritual community' – both foreign and domestic – in India. He described himself as 'too rational' and said that the people who prostrated themselves at temples acted 'very irrationally'.

References

Adam, B. (1994) *Time and Social Theory*. Cambridge: Polity Press.

Adam, B. (2004) *Time*. Cambridge: Polity Press.

Adler, J. (2002) The holy man as traveller: Early Christian asceticism and the moral problematic of modernity. In J. William, H. Swatos and L. Tomasi (eds) *From Medieval Pilgrimage to Religious Tourism: The Social and Cultural Economic of Piety* (pp. 25–50). Westport, CT: Praeger.

Arendt, H. (1977) *The Life of the Mind*. New York: Harcourt Brace Jovanovich.

Arellano, A. (2007) *Religion, Pilgrimage, Mobility and Immobility Religious Tourism and Pilgrim Management: An International Perspective*. Wallingford: CABI.

Aristotle and Jowett, B. (2004) *Politics*. Belle Fourche, SD: NuVision Publications.

Averroës and Arnzen, R. (2010) *On Aristotle's 'Metaphysics': An Annotated Translation of the So-called 'Epitome'*. Berlin: De Gruyter.

Badone, E. and Roseman, S. (2004) *Intersecting Journeys: The Anthropology of Pilgrimage and Tourism*. http://www.loc.gov/catdir/toc/ecip0414/2004002538.html

Bauman, Z. (1996) From pilgrim to tourist – Or a short history of identity. In S. Hall and P. du Gay (eds) *Questions of Cultural Identity* (pp. 18–36). London: Sage.

Bauman, Z. (2000) *Liquid Modernity*. Cambridge: Polity Press.

Bauman, Z. (2008) *The Art of Life*. Cambridge: Polity Press.

Beck, U. (2006) *Cosmopolitan Vision*. New York: Polity Press.

Behera, D.K. (1995) Pilgrimage: Some theoretical perspectives. In M. Jha (ed.) *Pilgrimage: Concepts, Themes, Issues and Methodology* (pp. 43–61). New Delhi: Inter-India Publications.

Bell, D. (1978) *The Cultural Contradictions of Capitalism*. New York: Basic Books.

Bhardwaj, S.M. (1973) *Hindu Places of Pilgrimage in India*. Berkeley: University of California Press.

Bittarello, M.B. (2006) Neopagan pilgrimages in the age of the internet: A life changing religious experience or an example of commodification? *Journal of Tourism and Cultural Change* 4 (2), 116–135.

Blackshaw, T. (2010) *Leisure*. London: Routledge.

Bourdieu, P. (2000) *Pascalian Meditations*. Cambridge: Polity Press.

Campo, J.E. (1998) American pilgrimage landscapes. *Annals of the American Academy of Political and Social Science* 558, 40–56.

Castells, M. (2004) *The Power of Identity, The Information Age: Economy, Society and Culture* (Vol. 2) (2nd edn). Cambridge, MA: Blackwell.

Castoriadis, C. (1998) *The Imaginary Institution of Society*. Cambridge, MA: MIT Press.

Clifford, J. (1997) *Routes: Travel and Translation in the Late Twentieth Century*. Cambridge, MA: Harvard University Press.

Cohen, E. (2004) *Contemporary Tourism: Diversity and Change*. Boston, MA: Elsevier.

Collins-Kreiner, N. (2010) Researching pilgrimage: Continuity and transformations. *Annals of Tourism Research* 37 (2), 440–456.

Dubisch, J. (2004) Heartland of America: Memory, motion and the reconstruction of history on a motorcycle pilgrimage. In S. Coleman and J. Eade (eds) *Reframing Pilgrimage: Cultures in Motion* (pp. 109–132). New York: Routledge.

Eliade, M. (1954) *The Myth of the Eternal Return, or, Cosmos and History*. Princeton, NJ: Princeton University Press.

Eliade, M. (1957) *The Sacred and the Profane*. New York: Harcourt, Brace and Company.

Eriksen, T.H. (2001) *Tyranny of the Moment: Fast and Slow Time in the Information Age*. London: Pluto Press.

Fabian, J. (2002) *Time and the Other: How Anthropology Makes its Object*. New York: Columbia University Press.

Frey, N. (1998) *Pilgrim Stories: On and Off the Road to Santiago*. London: University of California Press.

Giddens, A. (1990) *The Consequences of Modernity*. Stanford, CA: Stanford University Press.

Giddens, A. (1991) *Modernity and Self-identity: Self and Society in the Late Modern Age*. Stanford, CA: Stanford University Press.

Graburn, N. (1983) The anthropology of tourism. *Annals of Tourism Research* 10 (1), 9–33.

Guignon, C. and Aho, K. (2010) Phenomenological reflections on work and leisure in America. In M.R. Haney and A.D. Kline (eds) *The Value of Time and Leisure in a World of Work* (pp. 1–24). Plymouth: Lexington.

Hassan, R. (2008) *The Information Society*. Cambridge: Polity Press.

Hassan, R. (2009) *Empires of Speed: Time and the Acceleration of Politics and Society*. Leiden: Brill.

Heaphy, B. (2007) *Late Modernity and Social Change*. London: Routledge.

Heidegger, M., Macquarie, J. and Robinson, E. (2006) *Being and Time*. Oxford: Blackwell.

Holba, A.M. (2010) The question of philosophical leisure. In M.R. Haney and A.D. Kline (eds) *The Value of Time and Leisure in a World of Work* (pp. 39–57). Plymouth: Lexington Books.

Honoré, C. (2005) *In Praise of Slowness: Challenging the Cult of Speed*. San Francisco, CA: Harper.

Luhmann, N. (1982) *The Differentation of Society*. New York: Columbia University Press.

MacCannell, D. (1976) *The Tourist: A New Theory of the Leisure Class*. London: Macmillan.

Maoz, D. (2004) The conquerors and the settlers: Two groups of young Israeli backpackers in India. In G. Richards and J. Wilson (eds) *The Global Nomad* (pp. 109–122). Clevedon: Channel View Publications.

Marx, K., Milligan, M. and Struik, D.J. (1977) *Economic and Philosophic Manuscripts of 1844*. New York: International Publishers.

Mathieu, J. (2009) The sacralization of mountains in Europe during the modern age. *Mountain Research and Development* 26 (4), 343–349.

Melucci, A. (1998) Inner time and social time in a world of uncertainty. *Time and Society* 7 (2–3), 179–191.

Morinis, A. (ed.) (1992) *Sacred Journeys: The Anthropology of Pilgrimages*. Westport, CT: Greenwood Press.

Parkins, W. (2004) Out of time: Fast subjects and slow living. *Time and Society* 13 (2–3), 363–382.

Petrini, C. (2001) *Slow Food: The Case for Taste*. New York; Chichester: Columbia University Press.

Pieper, J. and Malsbary, G. (1998) *Leisure, the Basis of Culture*. South Bend, IN: St. Augustine's Press.

Preston, J. (1992) Spiritual magnetism: An organizing principle for the study of pilgrimage. In A. Morinis (ed.) *Sacred Journeys: The Anthropology of Pilgrimage* (pp. 31–43). Westport, CT: Greenwood Press.

Reader, I. (2007) Pilgrimage growth in the modern world: Meanings and implications. *Religion* 37 (3), 210–229.

Rountree, K. (2005) From Medieval pilgrimage to religious tourism: The social and cultural economics of piety. *Sociology of Religion* 66, 211–212.

Rousseau, J-J. and Butterworth, C.E. (1992) *The Reveries of the Solitary Walker*. Indianapolis: Hackett.

Rousseau, J-J., Stewart, P., and Vaché, J. (1997) *Julie, or, The New Heloise: Letters of Two Lovers who Live in a Small Town at the Foot of the Alps*. Hanover: University Press of New England [for] Dartmouth College.

Ruskin, J., Evans, J. and Whitehouse, J.H. (1956) *Diaries*. Oxford: Clarendon Press.

Simmel, G., Frisby, D. and Featherstone, M. (1997) *Simmel on Culture: Selected Writings* (pp. 174–186, 219–221). London: Sage Publications.

Singh, S. (2005) Secular pilgrimages and sacred tourism in the Indian Himalayas. *GeoJournal* 64, 215–223.

Slavin, S. (2003) Walking as spiritual practice: The pilgrimage to Santiago de Compostela. *Body and Society* 9 (1), 1–18.

Swatos, W.H. (ed.) (2006) *On the Road to Being There: Studies in Pilgrimage and Tourism in Late Modernity* (Vol. 12). Leiden: Koninklijke Brill NV.

Thoreau, H.D. (2007) *Walking*. Rockville, MD: Arc Manor.

Timothy, D.J. and Olsen, D.H. (eds) (2006) *Tourism, Religion and Spiritual Journeys*. New York: Routledge.

Tomasi, L. (2002) Homo viator: From pilgrimage to religious tourism via the Journey. In W.H. Swatos and L. Tomasi (eds) *From Medieval Pilgrimage to Religious Tourism*. Westport, CT: Praeger.

Turner, V. (1973) The center out there: Pilgrim's goal. *History of Religions* 12 (3), 191–230.

Turner, V. (1992) *Blazing the Trail: Way Marks in the Exploration of Symbols*. Tucson, AR: University of Arizona Press.

Turner, V. (2007) *The Ritual Process: Structure and Anti-Structure*. New Brunswick, NJ: Aldine Transaction (original work published 1969).

Turner, V. and Turner, E. (1978) *Image and Pilgrimage in Christian Culture*. Oxford: Blackwell.

Urry, J. (2000) *Sociology Beyond Societies*. Cambridge: Polity Press.

Urry, J. (2002) *The Tourist Gaze* (2nd edn). Cambridge: Cambridge University Press.

Urry, J. (2007) *Mobilities*. Cambridge: Polity Press.

van Loon, J. (1996) A cultural exploration of time. *Time and Society* 5 (1), 61–84.

Virilio, P. (1997) Speed and information: Cyberspace alarm!, accessed 25 October 2010. http://www.ctheory.net/articles.aspx?id=72 (¿)

Virilio, P. (2006a) *Original Accident*. Cambridge: Polity.

Virilio, P. (2006b) *Speed and Politics: An Essay on Dromology*. Los Angeles, CA: Semiotext(e).

Virilio, P. (2010) *The Futurism of the Instant: Stop-Eject*. Cambridge: Polity Press.

Weber, M. (2003) *The Protestant Ethic and the Spirit of Capitalism*. Mineola, NY: Dover Publications.

3 On the Periphery of Pleasure: Hedonics, Eudaimonics and Slow Travel

Kevin Moore

Slowness – as it intersects with travel, tourism and mobilities in general – reconfigures a persisting debate in the tourism literature. It does so by connecting that debate directly to two categories of current concern: environmental sustainability; and, personal and social well-being. In other words, the notion of 'slow' – and its philosophical assumptions – turns what has been largely a moralistic debate over the nature of modern, especially 'mass', tourism into a theoretically cast debate over the requirements for environmentally sustainable behaviours. Such behaviours are thought to enhance social, cultural and personal 'flourishing'. My argument is that 'fast tourism' was, and is, produced by a world that confuses and obscures questions of well-being in relation to travel. In this chapter I will explain how the integration of both hedonic and eudaimonistic approaches to well-being avoids this confusion and sheds light on the prospects for theorising 'slow travel'. Hedonic approaches emphasise the production of pleasure through the fulfilment of desires as the basis of well-being. Eudaimonic approaches, by contrast, are concerned with the achievement of well-being via a slower process of 'nature-fulfilment'.

The debates that I have referred to above were framed in 1975 when Louis Turner and John Ash's book *The Golden Hordes: International Tourism and the Pleasure Periphery* set the tone for an enduring debate over the ethical and moral qualities of tourist activity and tourism. We see this distinction reflected in popular discourse on the differences between the 'traveller' and the 'tourist'. Debate has focused on identifying the real or false nature of tourist/traveller experiences through the application of interlinked concepts such as 'authenticity', 'sincerity', 'hyper-reality', 'spectacle' and 'simulacra' (e.g. Bruner, 1994; Cohen, 1988; Eco, 1983; MacCannell, 1976; Taylor, 2001; Urry, 1990; Wang, 1999). It has also been the intellectual pivot point for various categories and typologies of tourism and tourist experiences that have sought to unravel the intricate threads of the institutional, ethical, moral and even existential tensions supposedly involved in being a tourist (e.g. Cohen, 1979; Smith, 1977).

Extending these concerns into the contemporary era we see the proliferation of labels for various forms of 'better' tourism, such as, 'sustainable tourism', 'responsible tourism', 'ecotourism' and 'ethical travel'. Scholarly journals, such as the *Journal of Sustainable Tourism*, as well as ethical tourist 'guides' such as Lucy Popescu's *The Good Tourist: An Ethical Traveller's Guide* (2008) are manifestations of discourses of concern over the nature and consequences of modern tourism (a point to which I will return). Advocacy for, and research into, slow travel are another turn of this conceptual spiral and, likewise, confront tensions that appear inherent in modern tourism. In fact, this debate goes well beyond considerations of travel and tourism. Over 30 years ago, MacCannell (1976: 9) related an anecdote involving an Iranian student of his in Paris that encapsulated both the emotional urgency of these tensions and the links between what is happening in tourism and in the wider world. At one point, the student came up to him and 'half shouting' said 'Let's face it, we are all tourists!' This comment, dripping with moral condemnation, was about the modern world with its apparent depthless, consumerist values. It was also an expression of some deep, ideological commitments to, and questions about, how we should live in a globalised world. As well as manifesting intellectual disdain for the 'democratisation' (i.e. massification) of tourism it also implicitly attacked increasing spatial mobility, economic inequality and social alienation and fragmentation.

In this light, a re-reading of Turner and Ash (1975) provides retrospective insights and historical, conceptual and ethical interest. They highlighted, for example, the term 'pleasure periphery' as a political and geographic relationship between tourist generating regions and destinations (Turner & Ash, 1975: 11–12). Beginning with the Mediterranean coast in Europe, and the Gulf and West coasts in America, the pleasure periphery inexorably spread outwards. Today it has gone global and, some would argue, virtual. The phrase 'pleasure periphery' orbits the question of the 'good life' and does so with a hedonic imperative. In almost classic hedonic terms, its focus is on the sensuous (and sensual) pleasures of 'sun, sand, sea and sex'. Significantly, the chapter (Chapter 6) in *The Golden Hordes* (Turner & Ash, 1975) entitled 'The Pleasure Periphery' is devoted to a discussion of the historical impact of jet aircraft in extending modern mass tourism into almost every nook and cranny of the post World War II globe. It is, of course, with the airliner that 'fast tourism' is truly born. The possibility of 'parachuting' tourists into remote locales and exotic cultures before being airlifted out some days or weeks later – brimful with pleasurable experiences – initially performed a 'piggy back' on the residual military airfields dotted around the world on islands, in inland wilds and in previously 'slowly connected' towns and cities in every country. Importantly, through this development, pleasure and 'speed' cleaved, almost inseparably, in the base assumptions of modern tourism and the modern mind. Tourism grew, that is, as a function of the 'compression' of global space and time relationships.

It is also worth remembering that, for Turner and Ash (1975), the question of the *speed* of travel and tourism (i.e. how fast or slow it is) was primarily a question about the rate of growth. In their final chapter on the future of tourism they pose the question of whether or not 'there are any larger [than the Cuban revolution] problems which could lead to the growth of international tourism being slowed down (or maybe reversed)?' (Turner & Ash, 1975: 280). That is, they assumed that modern tourism was just of this kind (i.e. 'fast') so the only escape from its negative consequences would be less tourism and fewer tourists, of which they could see little prospect. They could not have foreseen, of course, the concerns over climate change that exist today and which underpin much current interest in slow travel and feed into the slow movement in general, with its emphasis on locality. In response to the 1970s oil shocks, they did, however, speculate about the impact of an oil shortage on the jet 'fuelled' explosion in tourism. As costs of oil increased they asked '[d]oes this mean, however, that travel must now become so expensive as to slow down the growth of international tourism for good?' (Turner & Ash, 1975: 281). Wisely, they answered 'no', which, in hindsight, was largely correct.

Further, they spotted the trend towards an increasingly media-soaked world and suggestions that, in such a world, people would forego actually going places given that they could 'experience' places at a distance. Debates about the role and impact of social networking sites and technological connectivity notwithstanding, they once again, wisely opined that the 'trouble with this argument is that we have not yet bred people, other than Howard Hughes, who are willing to live vicariously through the media' (Turner & Ash, 1975: 283). If anything, virtual experiences of other places may reinforce the desire or even obligation to visit and gain an embodied experience of them. Any researcher interested in the prospects for 'slow travel' could do worse than begin with a slow (re-)reading of that chapter. With brutal realism they lay out the challenges for anyone hoping to reduce either tourism itself or, more modestly, its impacts.

In the following discussion, I want to pick up on a number of the points and themes raised in the above paragraphs. In particular, I want to consider the crossover between these ethical and moral dimensions of modern tourism and tourist activity, and the idea that tourism – at least for the individual tourist – is one part of the pursuit of well-being in the modern world. The desire to pursue the enhancement of one's well-being by having direct, embodied experiences of place via new mobilities is a double-edged sword. So long as travel and tourism are assumed to be opportunities for individuals to pursue a consumerist 'good life' through engaging in an endless and accelerating conveyor belt of activities, destinations and experiences, the prospects for slow travel making much of an impact on tourists' behaviours remain slight. If, however, travel and tourism are seen as part of a slower, gradually developed experience of the 'good life' and the pursuit of an integrated experience of well-being throughout one's life, slow travel can become a central ingredient, even assumption, of modern tourism.

To develop this argument, I will begin with a review of work on well-being that emphasises the integrated nature of so-called 'hedonic' and 'eudaimonic' approaches. The former places emphasis on desires and goals – that is, on results. The latter focuses on the process of living and notions of pleasure, satisfaction, etc., as peripheral (in relation to intention), but nevertheless important, by products of that process. Then I will apply the account of well-being that emerges from that review to the characteristics – and aims – of slow travel as it is both experienced and advocated. In passing, I will also note some tensions around the question of how the well-being of an individual might relate to the 'well-being' of the planet.

Peripheral Pleasure?

The literature on well-being (in its various guises as 'subjective well-being', 'happiness', 'positive psychology', 'wellness studies', 'the science of well-being', etc.) incorporates a definitional divide between hedonic and eudaimonic accounts. Simply, hedonic accounts focus on well-being as linked to the gaining of pleasure and the avoidance of pain. This is part of a long tradition from the Cyreniacs and Epicureans of ancient Greece to the Utilitarians of the 19th century. Eudaimonism, by contrast, highlights the centrality of flourishing, welfare and 'nature-fulfilment' and its concern can be summarised in Socrates' famous question: how ought one to live? (see Haybron, 2008). That question presupposed both the means and the freedom to contemplate and enact its answer. Of interest, of course, is that today those with such means and freedom are presented with the opportunity to enact the answer in either principally hedonic or eudaimonic ways, partly through involvement in the range of mobilities now available.

Theories of and broad approaches to well-being tend toward one or other of these accounts. The extensive work on 'subjective well-being' (SWB), with its focus on quantifiable measures of 'Positive Affect' (PA), 'Negative Affect' (NA) and 'Life Satisfaction' (LS) (e.g. Diener, 1984, 1994; Diener & Emmons, 1985; Diener et al., 1985; Eid & Larsen, 2008) fall under the hedonic heading. Rival, contemporary approaches linked to intrinsic motivation and notions of 'becoming', 'self-actualisation' and other humanistic psychological concepts, sit squarely in the eudaimonic camp (e.g. Deci & Ryan, 2008; Waterman, 1993). Research on SWB has principally focused on isolating the correlates of people's subjective perceptions of their well-being. The variables isolated range from genetic and personality-based traits through to marital status and friendship networks and, beyond, to national and global social and economic indicators (e.g. Gross Domestic Product (GDP), Gini coefficients, etc.). As Deci and Ryan (2008: 2) stated, '[i]n research on SWB, the primary focus has been on factors that lead to SWB – including person factors, social-environmental factors, and cultural factors'. Importantly, in this research, '[a]ssumptions have not been made about what should yield SWB nor about universality in the conditions that are likely to make people happy' (Deci & Ryan, 2008: 2). That is, the hedonic, SWB approach presupposes nothing about the conditions likely to produce well-being in humans.

Approaches to well-being, however, cannot ultimately be neatly compartmentalised in this way. Haybron (2008: 18), in a review of philosophical treatments of well-being, pointed out that '[n]one of the major schools of ancient ethical thought failed to maintain that the good life was a pleasant one'. That is, hedonic pleasure was seen as an outcome of living the 'good' or virtuous life: the hedonic and eudaimonic were fused. Further, 'in the Epicurean case [and for the Sceptics], the "static" pleasures of tranquillity or *ataraxia*' were considered well worth seeking (Haybron, 2008: 18). Stillness – 'slowness' – was understood as part of this fusion. More generally, it was the cultivation of the 'right' desires that was the focus of much discussion. Ordering one's inner emotional life would lead, even for the Stoics, to a pleasant, virtuous life characterised by *ataraxia*, *eupatheiai* ('good affects') 'including a kind of joy (*chara*)' (Haybron, 2008: 19). Aristotle famously argued in *The Nichomachean Ethics* that 'the virtuous life is also the most pleasant' (Haybron, 2008: 19). This version of well-being, that is, depended upon a much-debated psychology of well-being in which both pleasure and virtue were 'horse and carriage' components of the act of living well.

The relationship between pleasure and skill has long been acknowledged in leisure studies. While criticised for its focus on activity and achievement, Csikszentmihalyi's (1975; 1990) notion of the 'flow' experience supposedly emerges from the balance between learnt and practiced skills and the challenges of an activity. The pleasure, for example, gained from playing chess comes from playing opponents good enough to beat you sometimes but not so good that they always beat you. The more hedonistic end of this spectrum has also, however, often been seen as pivotal. The role of 'spectacle' (e.g. Urry, 1990) in travel is a clear example of this. Simply gazing upon spectacular views and constructions – or passive engagement in activities (e.g. everything from roller coasters to tandem skydiving) – provides the hedonic and status rewards (pleasure) with little requirement for the exercise of skill.

These different forms of leisure mirror a historical philosophical shift in which '[s]erious reflection on the psychology of well-being becomes relatively scarce, even as accounts of well-being grow ever more psychologised' (Haybron, 2008: 20). Haybron (2008) argues that this turnaround is directly related to a shift in views about 'personal authority' in matters of well-being. Ancients assumed that people generally did not know what was best for them. Even Epicureans assumed most people needed considerable education in their own interests. Indeed, according to Haybron (2008: 20), 'there may have been no domain of personal welfare in which ancient philosophers considered the typical person to be authoritative' and, thus, 'the standard economic view of modernity – that well-being consists roughly in people getting whatever they happen to want – would have seemed childish if not insane to most ancient thinkers'.

By contrast, for Enlightenment and post-Enlightenment philosophers, the prime 'good' was freedom which, when combined with the belief that individuals are best situated to make judgements about their own well-being, led to the – curiously ironic – conclusion that what people need is not 'enlightenment' but 'empowerment'. The latter, of course, is to be delivered via economic freedom – the freedom of individuals to gain the resources necessary to pursue their self-set, well-being goals. In Isaiah Berlin's famous typology (Berlin, 1958), it is the *negative freedom to* do as one chooses without the constraint of others, versus the *positive freedom from* limitations inherent in one's circumstances (e.g. poverty, lack of knowledge, education or a skilled rational faculty).

Current research on SWB, despite its focus on hedonic components of well-being, has begun to redress this pragmatic, economic bias in understanding well-being. This is not surprising. Any causal account of well-being will tend to seek to isolate its determinants. 'Nature-fulfilment', for example, is central to eudaimonic approaches to well-being. Whether it is the Maslowian directive to 'become what you are' (Maslow, 1987) or general guidance to exercise capacities, talents and skills to one's potential, the idea that the good life involves being whom – and what – you are has enjoyed enduring appeal. As already mentioned, travel itself has been said to be driven by notions of 'authenticity' and 'sincerity' (MacCannell, 1976; Taylor, 2001). More directly, research on SWB has also undermined the assumption that well-being primarily depends on economic factors. Much-publicised findings (Easterlin, 1995) that levels of SWB seem independent of increases in per capita income over a low (by developed nation standards) threshold have encouraged the conviction that absolute levels of income or wealth are independent of well-being. There have also been calls for countries to abandon or supplement economic indicators of 'well-being' (GDP, Gross National Product (GNP), etc.) with indicators linked directly to SWB (see Diener & Seligman, 2004).

The research is, however, more nuanced than this (Diener, 2008). SWB within a nation, for example, is higher the wealthier someone is, on average. The latest global survey (Diener & Ng, 2010) has also revealed that Life Satisfaction is consistently higher the higher a nation's GDP. That same survey, however, also found that Positive Affect levels were unrelated to wealth and depended, as much research has shown, on the quality of social connections. This focus on 'connections' raises important, but so far largely unanswered, questions about the ways in which mobilities affect well-being.

The following section explores some of these potential relationships between well-being and slow travel.

Slow Travel, Slowness and Hedonism

Dickinson and Lumsdon (2010) note that any definition of slow travel should regard both transportation and experiential components of travel. They note that slow travel is a conceptual framework that involves people who 'travel to destinations more slowly overland, stay longer and travel less' and who incorporate travel to a destination as itself an experience and, once at the destination, engage with local transport options and 'slow food and beverages', take time to explore local history and culture and 'support the environment' (Dickinson *et al.*, 2010a&b cited in Dickinson & Lumsdon, 2010: 1–2). The first thing to notice about these definitions is how well they connect with the debate over the ethics of tourism I have previously mentioned. Slow travel and tourism, in that context, is a 'better' form of travel and has close family resemblances with other supposed forms of 'alternative tourism' such as 'ecotourism', 'responsible tourism' and 'ethical tourism'. It is a kind of tourism that, according to its advocates, people *should* engage in. As Dickinson and Lumsdon (2010: 75) explicitly state, '[t]he challenge for tourism in the 21st century is seemingly how to reshape itself so that people can continue to enjoy their leisure time, while, at the same time, the supply sector manages to avoid the worst scenarios of climate change'.

The second thing to notice is that the definitions of slow travel assume that it is – or can be – enjoyable, and is so in ways that 'fast tourism' is not. There is an assumption that 'people can continue to enjoy their leisure time [and travel experiences]' and save the planet at the same time. This is a revealing commitment. While Turner and Ash (1975) advocated *less* tourism, slow travel advocates seem not to want to deny people travel but to 'nudge' them into more environmentally (and, presumably, socially) 'friendly' ways. Well-being – understood eudaimonistically – intersects with longer and 'deeper' social and physical engagement with places and this indeed provides some hope that the individual well-being derived from 'slow travel' might intersect with environmentally sensitive forms of travel. Tourism that involves less distance but more intensive engagement (perhaps through public transport options, cycling, etc.) should also support well-being, under-stood less as 'pure' consumer hedonism. Like many attempts to advocate for less harmful travel and tourism it encounters obstinacy on the part of travellers and tourists. Slow travel is a 'promising niche' (Matos, 2004: 96) or a 'niche activity' (Dickinson & Lumsdon, 2010: 103). There is hope that, as tourism's role in climate change impacts and 'becomes more widely recog-nized, it is likely that more consumers will seek to change their ways'

(Dickinson & Lumsdon, 2010: 103). Yet, even if individual well-being can be served by slow travel in ways that hedonistic travel cannot, a significant problem remains. The interests of powerful industry institutions are acknowledged as perpetuators of 'fast tourism', as are an array of social representations, practices and discourses (see Dickinson & Lumsdon, 2010: 47–74). This acknowledgement perhaps explains the sense of hopeful desperation that surrounds slow travel. 'Hopeful' because it does offer a viable option that might ameliorate or even eliminate many adverse consequences of modern tourism. 'Desperation' because it seems quite possible, even likely, that the consequences will continue to accrue just because 'fast tourism' is entrenched at so many levels.

How has it come to this? It is well documented that modern society has become faster as its social institutions (e.g. school and the workplace) have become more time regulated and 'accelerated'. Less well acknowledged, is that leisure lives have also suffered the same fate (e.g. Rojek, 2010; Ryan, 1997; Urry, 1994). Woehler (2004) claimed that, as a result of the time-ordering of work and economic life, 'personal time' arose in the sense of individuals 'owning' their own time. Since work provides a fundamental and inescapable ground for modern experience, leisure, in turn, becomes heavily time-structured. It also becomes the site of our ability (and requirement) to write our own biography (Beck, 1986) in a world in which traditional environments that channel personal development (and destiny) are lacking. Leisure thus becomes an arena manifesting the *obligation* to specialise in one's life project. While at leisure in the modern world, that obligation confronts '[b]oundless possibilities of action [which] lead to a diversity of leisure lifestyles with different leisure time activities' (Woehler, 2004: 87). In tourism, these leisure options produce a 'multioptional vacation landscape'. But, since tourists remain uncertain as to what options will deliver the obligatory fulfilment and 'their "real" identity, pressure increases to constantly realize different or additional activities and experiences' (Woehler, 2004: 89). Out of this pressure cooker 'fast leisure' and 'fast tourism' are born. It is this anxiety that leads many tourists to opt for package tours and holidays to alleviate the feelings of a lack of time in the midst of the pressure for (identity) fulfilment. Speed and relinquishing decision-making becomes the way to fulfil the obligation through the path of least resistance.

As Rojek (2010: 1–2) argues, leisure cannot now be equated to time off work. The equation that sees less work equal more leisure and, then, more leisure equalling more freedom no longer holds in a world in which leisure requires increasing consumption and, hence, increasing income (and work). Connecting with the obligation to write one's own biography, Rojek (2010: 3) sees leisure as where considerable 'emotional intelligence' is required (to fulfil the (post)modern imperatives to at least be seen to express 'care for the self' and 'care for the other'). It involves a disciplined employment of 'emotional labour' to acquire this intelligence – a constant monitoring of one's performance and of the environment to detect clues for the emotionally correct responses.

Referring to the ancient Greeks, Rojek (2010: 189) notes that Aristotle 'understood that to be an epicure of leisure requires discipline and preparation' and '[t]o truly enjoy leisure we must harness ludic discipline to gain wisdom from scarcity'. Leisure is a 'school for life' and '[t]he end of schooling is to maintain and enhance competence, relevance and credibility' (Rojek, 2010: 189). For Rojek, leisure is now imbued with a refined, social-ethical, emotional labour. Whether or not he is right hinges on whether or not the ancient philosophers were right: that hedonics and eudaimonics are intimately interwoven. The appeal of slow travel as an environmentally responsive and culturally engaged mode of travel also depends on this interweaving as its hope must be that it incorporates the pleasure that arises from this slow and appreciative immersion in tourist experience. Cycling around a city or a region, for example, engages a multiplicity of sensory and social experiences of an environment and leads to the kinds of satisfaction, and pleasures, that can only emerge from such engagement.

Conclusion

The prospects for slow travel mirror the prospects for modern life. Human well-being, whether at the individual or global-historical level, is currently askew. An overly materialist and consumerist life is not, typically, an especially happy one (see Kasser, 2002). Yet, despite this knowledge, most people continue to engage in such a life. King (2008) has pointed out that developmental approaches often have a 'maturity value'. They assume the 'rightness' of certain developmental outcomes over others. Further, one 'surprising' outcome of the pursuit of happiness is to assist in the development of 'maturity', or what she calls 'wisdom', understood as 'the fundamental pragmatics of life permitting exceptional insight, judgment, and advice involving complex and uncertain matters of the human condition' (Baltes & Staudinger, 1993: 76, cited in King, 2008: 441).

The issue for advocates of 'friendlier' travel is how to encourage travellers to become 'wise', both in relation to their own well-being and that of the planet. There is, however, no necessary connection between the pursuit by individual tourists of well-being and the 'good' of the planet. Slow travel remains only a 'niche' because it reflects an enduring – and widely institutionalised – misunderstanding of the relations between leisure, well-being and human freedom. Becoming autonomous, self-governing individuals capable of maximising well-being requires considerable effort and 'discipline', what King (2008) refers to as 'wisdom' in relation to life's practicalities. Also, the characteristics of slow travel with its more human-scale speed and immersion (see Dickinson & Lumsdon, 2010; Matos, 2004) reflect characteristics required for the development of an integrated human well-being: close social relationships; absorption of the features of environments and the steady practice of, and improvement in, skills and knowledge of people, places and mobilities. Such well-being combines both hedonic and eudaimonic components as complementary features.

The emergence of slow travel is part of a broader reclamation of a more complete understanding of our well-being. Slow tourism also reaffirms tourism as a significant 'channel' for, and expression of, the tensions people face in today's world. Further, it demonstrates how some people might weave an awareness of the globalised world and its problems into their efforts to live life well and pleasurably. In this way, individual well-being can become linked to – perhaps even dependent upon – global well-being. Slow travel and tourism provides an opportunity for this linkage. Turner and Ash (1975) railed against a tourism that, they believed, desecrated all it touched and that was an ugly manifestation of an ugly world. They spoke of the 'centre' and its 'periphery of pleasure' with, perhaps, unnoticed irony that they were also assuming that hedonistic 'pleasure' was far from peripheral but was at the centre of modernity and life's strivings. Yet, pleasure cannot sustain itself at the centre of life: to maintain itself, pleasure must be on the periphery. Pleasure is best approached indirectly. Once the process of living is mastered, pleasure follows in its rightful place, as consequence rather than goal. Slow travel and tourism should promote – rather than apologise for – the fact that its practice is on the periphery of pleasure.

References

Beck, U. (1986) *The Risk Society: Towards a New Modernity*. London: Sage.
Berlin, I. (1958) *Two Concepts of Liberty: An Inaugural Lecture Delivered Before the University of Oxford on 31 October 1958*. Oxford: Clarendon Press.
Bruner, E.M. (1994) Abraham Lincoln as authentic reproduction: A critique of postmodernism. *American Anthropologist* 96 (2), 397–415.
Cohen, E. (1979) A phenomenology of tourist experiences. *Sociology* 13, 179–201.
Cohen, E. (1988) Traditions on the qualitative sociology of tourism. *Annals of Tourism Research* 15, 29–46.
Csikszentmihalyi, M. (1975) *Beyond Boredom and Anxiety*. San Francisco, CA: Jossey-Bass Publishers.
Csikszentmihalyi, M. (1990) *Flow: The Psychology of Optimal Experience*. New York: Harper and Row.
Deci, E.L. and Ryan, R.M. (2008) Hedonia, eudaimonia and well-being: An introduction. *Journal of Happiness Studies* 9, 1–11.
Dickinson, J. and Lumsdon, L. (2010) *Slow Travel and Tourism*. London: Earthscan.
Diener, E. (1984) Subjective well-being. *Psychological Bulletin* 95, 542–575.
Diener, E. (1994) Assessing subjective well-being: Progress and opportunities. *Social Indicators Research* 31, 103–157.

Diener, E. (2008) Myths in the science of happiness, and directions for future research. In M. Eid and R. Larsen (eds) *The Science of Subjective Well–Being* (pp. 493–514). New York: Guilford Press.

Diener, E. and Emmons, R.A. (1985) The independence of positive and negative affect. *Journal of Personality and Social Psychology* 47, 1105–1117.

Diener, E. and Ng, W. (2010) Wealth and happiness across the world: Material prosperity predicts life evaluation, whereas psychosocial prosperity predicts positive feeling. *Journal of Personality and Social Psychology* 99 (1), 52–61.

Diener, E. and Seligman, M.E.P. (2004) Beyond money: Toward an economy of well-being. *Psychological Science in the Public Interest* 5 (1), 1–31.

Diener, E., Emmons, R.A., Larsen, R.J. and Griffin, S. (1985) The satisfaction with life scale. *Journal of Personality Assessment* 49, 71–75.

Easterlin, R.A. (1995) Will raising the incomes of all increase the happiness of all? *Journal of Economic Behavior and Organization* 27, 35–38.

Eco, U. (1983) *Travels in Hyperreality*. San Diego, CA: Harcourt Brace Jovanovitch.

Eid, M. and Larsen, R.J. (2008) *The Science of Subjective Well-Being*. New York: Guilford Press.

Haybron, D.M. (2008) Philosophy and the science of subjective well-being. In M. Eid and R. Larsen (eds) *The Science of Subjective Well-Being* (pp. 17–43). New York: Guilford Press.

Kasser, T. (2002) *The High Price of Materialism*. Cambridge, MA: MI1 Press.

King, L.A. (2008) Interventions for enhancing subjective well-being: Can we make people happier, and should we? In M. Eid and R. Larsen (eds) *The Science of Subjective Well-Being* (pp. 431–448). New York: Guildford Press,

MacCannell, D. (1976) *The Tourist: A New Theory of the Leisure Class*. New York: Schocken Books.

Maslow, A.H. (1987) *Motivation and Personality*. New York: Harper and Row.

Matos, R. (2004) Can slow tourism bring new life to alpine regions? In K. Weiermair and C. Mathies (eds) *The Tourism and Leisure Industry: Shaping the Future* (pp. 93–103). New York: The Haworth Hospitality Press.

Popescu, L. (2008) *The Good Tourist – An Ethical Traveller's Guide*. London: Arcadia.

Rojek, C. (2010) *The Labour of Leisure: The Culture of Free Time*. Los Angeles, CA: Sage.

Ryan, C. (1997) 'The time of our lives' or time for our lives: An examination of time in holiday. In C. Ryan (ed.) *The Tourist Experience: A New Introduction* (pp. 194–205). London: Cassel.

Smith, V.L. (1977) *Hosts and Guests: The Anthropology of Tourism*. Philadelphia: University of Pennsylvania Press.

Taylor, J.P. (2001) Authenticity and sincerity in tourism. *Annals of Tourism Research* 28 (1), 7–26.

Turner, L. and Ash, J. (1975) *The Golden Hordes: International Tourism and the Pleasure Periphery*. London: Constable.

Urry, J. (1990) *The Tourist Gaze: Leisure and Travel in Contemporary Societies*. London; Newbury Park, CA: Sage.

Urry, J. (1994) Time, leisure and social identity. *Time and Society* 3, 131–151.

Wang, N. (1999) Rethinking authenticity in tourism experience. *Annals of Tourism Research* 26 (2), 349–370.

Waterman, A.S. (1993) Two conceptions of happiness: Contrasts of personal expressiveness (eudaemonia) and hedonic enjoyment. *Journal of Personality and Social Psychology* 64, 678–691.

Woehler, K. (2004) The rediscovery of slowness, or leisure time as one's own and as self-aggrandizement? In K. Weiermair and C. Mathies (eds) *The Tourism and Leisure Industry: Shaping the Future* (pp. 83–92). New York: The Haworth Hospitality Press.

4 Slow'n Down the Town to Let Nature Grow: Ecotourism, Social Justice and Sustainability

Stephen Wearing, Michael Wearing and Matthew McDonald

This chapter develops an account of *slow* ecotourism that challenges objectifying and commodifying processes in local communities of both the developing and developed world. We argue for sustainable forms of slow ecotourism that promote social justice for host communities and high standards of environmental protection as a basis for the economic, social and environmental protection of these communities. Models and practices for tourist enterprises are suggested in later sections of this chapter that value local workforce participation and equity for host workers and their communities. We argue that 'slow ecotourism' needs to embrace principles of both social justice and sustainability to achieve this. The adoption of these principles would ensure that Indigenous and host communities receive equitable and positive redistributive socio-economic effects and poverty alleviation through the creation of jobs, as well as a fair share of the profits that may accrue from ecotourism. The focus on decommodification allows for anti-poverty strategies via alternative forms of exchange that are slow, informal and bartered.

One component that could assist in achieving some of the goals stated above is in development that is *slow*. Slow development occurs at a pace that communities are able to deal with. Development comes in all shapes and sizes; however our focus is on ecotourism more specifically, either as a stand-alone programme or as a part of a mixture of programmes to help alleviate poverty and its associated problems. Ecotourism's global following stems from its perceived potential to deliver benefits to communities remote from large centres of commerce that do not involve widespread social or environmental destruction. Too often in the past, the only choices available for remote communities were extractive industries – mining, logging, or fishing – which in many cases have led to deleterious impacts on local communities and which have often left an unacceptable legacy of long-term environmental damage. Fundamental to the relationship between ecotourism and tourism in general, is the return of revenues back to resource protection and to the local community itself. Much of the support for ecotourism is based on the belief that it will generate funds above and beyond those needed to manage the visitation, but as yet no framework has been comprehensively developed to assess these issues. One common sacrifice that communities have made in the past has been to allow the increase in development to a level that changes the pace of those communities.

To understand this issue in more depth it is important to review the research and practical applications that have led to ecotourism acting as an agent for positive change in communities based on the ideas of 'slow'n down the town', particularly in rural and regional areas that define themselves by their relative slowness in comparison to large cities. The predominance of modern Western business approaches within tourism has tended to exclude communities and their desire to maintain a slow pace of life. Similarly, it is often the desire of the tourist to enter and enjoy this pace of life as a part of the holiday they wish to experience. Slow tourism allows for a different set of exchanges and interactions than those available in the hurried contexts of mainstream tourism and the conventional understanding of economic benefits to the host or cultural benefits to the tourist. Dickinson and Lumsdon (2010: 4) outline themes across the literature that define slow tourism, suggesting slowness equates to quality time (a physical slowing down of time for enjoyment to occur) and a quality experience that allows engagement with 'other' and 'nature'. This in effect combines slow principles of travel with those of authentic eco-tourist experience.

Slow ecotourism has the potential to establish and sustain links between conservation, the tourism industry, communities and nature. Social and environmental benefits can accrue as a result of ecotourism increasing overall standards of living due to the localised economic stimulus provided by increased visitation to the site. Environmental benefits accrue as host communities are persuaded to keep local cultural elements and ecological spaces and places in order to sustain economically viable tourism. The exploration of these links is fundamental to our analysis in its attempt to establish a conceptual framework, which supports 'slow'n down' as a mode of experience in tourism. The 'slow' movement seeks to resist mainstream rationalities and standardisation of tourism. Such resistance and reframing into a slow ecotourism can counter the time-space compression of globalisation and the global business of tourism, whilst also providing robust conditions for local touristic economies to thrive and become sustainable. The Slow Food movement, for example, is recognised for its significant economic impacts on rural and regional areas through the development of alternative food distribution networks, where food is sourced from local and small-scale producers (Parkins & Craig, 2006).

Conceptualising Slow Ecotourism

The conceptual framework for this chapter will also rely upon several key authors in contemporary social theory and tourism. The use of the terms commodification and de-commodification are accepted concepts in comparative politics and social policy understandings of differences in Western welfare regimes (Hicks & Esping-Andersen, 2005). The regimes of welfare capitalism allocate disposable time and clearly constitute different welfare cultures that are more or less generous or stringent in their delivery of social welfare goods and services and in orchestrating redistribution (Goodin *et al.*, 2008). We also use the term commodification to connote what Simmel (1991) argued was the basis of the money economy in capitalism, and what has been more recently framed as the commodification of emotions (Hochschild, 2003). Money, Simmel (1991: 17–18) observed, mediates all relations, creating a separation and distance between the individual and their social, material and natural worlds. The money economy is predicated on free and open markets, based on a theory that economic growth is most rapid when goods, services and capital can circulate without let up or hindrance (MacEwan, 1999: 31; Mandel & Novack, 1970: 76). Embedded in the capitalist economic system is money (or capital) which acquires its meaning by being 'given away' or 'circulating'. The unceasing movement of capital around the global market speeds all related activities and makes them continuous; the freer the market, the faster money can flow (Simmel, 1990).

In order for the local economies of communities to protect themselves from the excesses of competition in the global market they need to rely upon their own labour and resources to build sustainable slow tourist products and destinations. Local communities need to control and self regulate their tourist industries, and we are suggesting that slow ecotourism offers the best options in terms of environmental protection and long-term, sustainable economic growth in these communities. As Adam (2004: 39) states,

> When time is money, then the production of something of equal quality in a shorter time allows for a reduction in the price of the product, which increases its competitiveness. Equally, the faster an invention comes to market the better for the competitive edge over business rivals. To be first, that is, faster than competitors, is crucial, and this applies whether the 'product' is a new invention, garment, news story or a new drug. Thus, when time is money, speed becomes an absolute and unassailable imperative in business.

Some authors in the area of ecotourism (Kosoy & Corbera, 2010) have rethought how commodity fetishism, following Marx, can impact on local communities. The overall effect of commodification of ecotourism in developed and developing capitalist market economies is that human beings, social relations and the natural environment are viewed and treated as 'things' (or commodities) to be packaged, presented and sold in the marketplace quickly and in great quantity, with little worth being ascribed to their intrinsic value.

Ecotourism constitutes approximately 3–4% of global tourism depending on the broadness of definition (World Tourism Organisation, 2002). After considering a range of definitions Fennell (2003: 25) provides an integrated understanding of the term: 'Ecotourism is a sustainable form of natural resource based tourism that focuses on experiencing and learning about nature, and which is ethically managed to be low budget, non-consumptive and locally oriented.'

Already in this definition we can see the appeal and potential for travellers to experience, in-depth, a slower mode of activity in ecotourism. This definition also indicates the significance of being embedded in local contexts that can enthral and engage the traveller. Such a perspective offers potential for a less commodified and marketised form of tourism. From this conceptual framework we can then ask 'what is the nature of slow ecotourism in the emerging global economy?' The constituent qualitative elements are time, activity and experience. Slow ecotourism has the potential to change our perceptions and attitudes towards nature and the use of land and environment as scarce resources that need to be nourished and nurtured, as opposed to being exploited. Slow ecotourism is distinguished from brief and commodified ecotourism experiences ('been there and got the T-shirt'), as distinct from the embodied memory of living time and space with greater attention focussed upon the self-world/other culture relationship.

Such a view is shared with international nature advocates such as the World Wildlife Fund (WWF), part of whose underlying vision is to 'seek to instil in people everywhere a discriminating, yet unabashed, reverence for nature and to balance that reverence with a profound belief in human possibilities' (World Wildlife Fund, 2010). This is not to romanticise the existing social and economic conditions and inequality in either developing or developed countries, but to look closely at what has worked in these societies in terms of appropriate ecotourism development and their interaction with all cultural constructions of nature. Can slow ecotourism strategies become part of this appropriate development in the global North and South? In a technological sense, sometimes slower development itself can force a slower sensibility, pace and timing of travel. For example, the use of cars designed in the 1950s from the USA and UK in countries such as Cuba (utilising Russian parts) and India (based on British designs) are interesting cases that have literally wound back the technological clock in these countries to enable the middle classes to buy affordable cars. Cuba survived the peak of oil crisis from the 1970s to today through such measures as the adoption of slow lifestyles, utilising cycling, encouraging walking and the use of public transport and community gardens. Nonetheless, as a private mode of transport, the carbon footprint for cars per capita is very high. So, the slow tourist movement would support tours that are oriented towards public transport as well as travel providers that use trains, buses, push bikes or walking as their mode of transport in and around local communities.

The slowing of the pace of industrial and global economic change can be one strategic answer to the encouragement of a sustainable tourism that acts in concert with a slowing down of the economy. Decommodification can then be linked to de-marketisation of ecotourism and a slowing down of social relationships at a local economic level. Boulanger (2009) has usefully suggested that rich Western societies need to adopt three strategies to transition to sustainable consumption that includes leisure and tourist consumption. Core to this transition is the strategy of decommodification in line with strategies of eco-efficiency and sufficiency for communities and local markets. There are two kinds of decommodification processes, those that move goods and services from the market sphere to the community sphere, or the informal and formal networks that constitute civil society, and those that move from the market to the public sphere under the watch of government and powerful political actors and interests. The latter is a strategy adopted in Australia in relation to solar power that is cross subsidised by the Federal Government and which provides offsets for private providers to individual households. While not a fully decommodified programme it leaves open future possibilities for transition to non-marketised and sustainable solar to electric power supply for households. The environmental and financial practicalities of using solar power as an incentive via the public sphere can also be harnessed for use in ecotourism and slow travel, where the aim is to cut back on the overuse of electricity and deliver non-coal fired sources to local communities. Boulanger links broad decommodifying strategies to providing a basic income structure in communities that allow citizens to maintain a modicum of purchasing power, 'decommodified consumption practices such as pooling, sharing, lending, and bartering result in more people using the same thing so that less of it has to be produced' (Boulanger, 2009: 6). Another clear example given by Boulanger (2009: 7) in rich nations are the Local Exchange and Trade Systems (LETs) that exist in Australia, Canada and the US that create sustainability by contributing to 'the lessening of the ecological footprint of consumption by relocating the economy, by decreasing the transportation costs and pollution, and through fostering sharing, pooling, recycling and repairing'. Nonetheless community-based forms of provision can create other human and social costs. There is a need to consider the ethical dimensions raised by fair trade and the eco-strategies of sustainability for both humans and nature (see Zaccai, 2007).

Embedding Social Value

The initial treatment of tourism destinations in developing countries typically made implicit assumptions that 'locals' were pre-modern, primitive, poor and technologically backward, while their (Western) 'guests' were modern, sophisticated, wealthy and technologically advanced. However, this binary classification has gradually faded away as many local communities in developing countries are looking beyond the blight of mass tourism to focus on the possible benefits of smaller-scaled, community-based tourism projects (e.g. Aramberri, 2001; Chan, 2006; Cole, 2007; Lyons & Wearing, 2008; Mbaiwa, 2004; Meethan, 2001; Milne & Ateljevic, 2001; Sherlock, 2001; van der Duim et al., 2005; Wearing & McDonald, 2002). It is suggested that slow tourism can reconfigure the tourist destination as an interactive space where tourists become creative actors engaging in behaviours that are mutually beneficial to local communities and where community members become agents in changes to their own local economy rather than 'acted upon' by outside commercial and governmental interests.

Local communities, particularly those in remote and rural locations around the world, are looking to improve their conditions by instituting tourism development where significant economic benefits transpire (e.g. Williams & Shaw, 1999). For less developed countries facing declining terms of trade for agricultural products and protectionist policies in the West, tourism is seen as an alternative route to economic growth (Sinclair, 1998). Along with general economic expansion, tourism has enabled some workers to move from lower paid jobs in the informal and domestic sector to higher paid jobs in the formal sector. Nonetheless, 'tourism is characterised by high growth and, with the exception of the airline sector, low protectionism' (Sinclair, 1998: 38).

As alluded to by Sinclair (1998), a number of problems arise with the use of developing countries' environmental resources for tourism; the most significant of these from an economic perspective is 'market failure'. In virtually all elements of the tourism industry, developing countries are unable to compete with services from the developed world such as airlines, hotels, travel agents and tour guides. Hoteliers in developing countries, for example, often lack the knowledge and/or skills required to successfully negotiate with international hotel management companies and tour operators. In countries with less human capital and workforce capacity the lower levels of negotiating skill contribute low returns per tourist to local economies (Sinclair, 1998: 39). In cases like this, where local communities are unable to compete, their participation in the tourism industry withers, resulting in the lion's share of tourism income being taken away or 'leaked' out from the destination (Liu, 2003). In this process of supposed 'tourism development', local communities and their environmental resources are commodified by outside agents. In effect, developing countries are subjected to a process of rationalisation inherent in the neoliberal economic system.

Therefore, there is a need to examine how a slower model or approach to tourism that can counteract these commodifying processes could be developed so that the relationship between local cultures and tourists is actively repositioned. The 'othering' process positions developing countries as inferior and is a sharp example of how local and Indigenous people are objectified in order to sell tourist experiences. One approach to re-orientate this relationship is the concept of 'social value', which, in the context of tourism, seeks to endorse local people and cultures. The idea is to create a tourism space where local communities play a central role in the planning and management of tourism in the places where they live. As a part of this process, micro-social elements need to be analysed because these are fundamental to the conceptualisation of tourist destinations. For example, how is the life of a community member improved through tourism; do they go from running a business to working in a resort as a manager; is this a better quality of life for them? This emphasis is often overlooked in the sociological analysis of the tourist experience, where the focus instead is typically on macro-social influences, impacts of tourism upon destinations, the quality of the tourist experience, and industry construction of the experience.

Social value is created through the way tourists and locals interact in the tourist destination. Ideally, tourists take their meaning of the site from the people who occupy it. The interactive dimension of the site represents a social process where a place has significance for the people who occupy it and the tourists who visit it. Cunningham (2006) argues that social valuing of the visited place can both enhance the tourist experience and enrich the culture and identity of the local population. He presents a case study of the Japanese island of Ogasawara, where local cultures and heritage are greatly undervalued by tourists and the tourism industry. In order to reverse this trend, Cunningham argues that the 'Obeikei' community could better communicate to visitors their unique understanding of, and value for, the place that is their island – its natural resources, remoteness and rich cultural history. Cunningham (2006) suggests the local community find ways of describing and representing their unique identity as 'islanders' to eco-tourists. Exposure to messages of local value would enable tourists to engage with the island's history at the invitation of the locals on their terms. The result would be a broadening of the tourism experience of both the local and the tourist. The locals might find that their culture and local identity are affirmed, while the traveller would have a more meaningful experience engaging with local knowledge and understanding. As Taylor (2001: 16) notes, 'important local values are promoted through tourist–local interaction, communication and engagement with the local'.

When locals are given a voice in the tourism development process, they are given an opportunity to communicate the social value of their places.

In other words, messages have the potential to be presented to tourists that provide an important point of interest and empathy for local communities (Cole, 2007). However, in instances where locals are positioned by the tourism industry as being at the bottom of the tourism hierarchy, meaningful interaction between them and tourists is difficult. The tourism experience for both locals and tourists is thus lessened as a result. If local communities are motivated and supported to represent their position in the tourism hierarchy, then there is potential for them to identify, clarify and advocate their valuing of place and, subsequently, for tourists, to experience the place and the way of life of local cultures.

Slowness, Harmony and Sustainability

While tourism is typically viewed as a negative force in many local communities there are ways that the potential problems associated with it can be ameliorated. With the right approach to participation and planning, tourism has the potential to act as a tool for sustainable community development and poverty reduction (Beeton, 2006). Slow ecotourism can create a new sense of time, place and experience in local communities. For example, slow travel can create for tourist and host alike a different and potentially deeper set of experiences in contrast to the fleeting and highly commodified tourism that is packaged by many travel agents.

Manyara and Jones (2007) studied six tourism-focused, community-based initiatives in Kenya by carrying out interviews with community leaders, managers, academics, support organisations, government officials and community members. Their findings indicate that potential benefits from such initiatives are proportional to the level of community involvement – the higher the community involvement the greater the benefits.

> Local communities and their leaders, for instance, need to be adequately sensitised and empowered so that they can make informed decisions to enhance sustainability and to secure appropriate capacity building to enhance skills and knowledge and promote transparency. Moreover, an appropriate policy framework is crucial for guiding CBE developments. The policy framework should address partnership and land ownership issues. (Manyara & Jones, 2007: 641)

In another example, Al-Oun and Al-Homoud (2008) investigated the potential for tourism to stem population displacement as a result of desertification in the Badia Desert in Jordon. In remote and rural environments, where people continue to live in traditional ways, the authors argue that a community-based approach to tourism is likely to be most beneficial for the local communities. The proposed tourism venture for the area was developed by carrying out extensive research in the initial phases comprising field interviews, field surveys, archival research and a pilot tourism project. The findings indicate that the success of this model depends on community development and control, an appraisal of the unique tourism resources in the area, a deep knowledge of the social values of the locals, and the creation of non-governmental organisations (NGOs) and partnerships with government agencies to enable communities to work together in order to transcend tribal differences. Of course, all of this takes time, so that slow ecotourism often depends for its sustainability on slow and considered development.

It is important, therefore, that local communities and their unique social values are understood by those seeking to develop tourism in a particular local community, and if they so desire, communicated to interested tourists. There is also the question of protecting social values and the impact that outside tourists might have on these. There is the need here for a sharp and penetrating analysis of values and localised ethos that reiterates the activism laid out by the US academic Deborah McLaren (1998) amongst several others in Australia and Canada (Wearing & Neil, 2009; Wearing & Wearing, 1999; Fennell, 2003). McLaren argues for collaborative North and South participant tourism strategies of local decommodifying ventures in ecotourism. Her words ring true today in her advocacy for a green, just and sustainable ethos for slow travel and ecotourism:

> Those of us in the North who reject the advance of commercialised global culture and those from the South victimised by it voraciously oppose the continued devastation of the environment and Indigenous populations. We need to take a hard look at the travel industry, at the self-exploitation of communities, and the roles we play as individuals. (McLaren, 1998: 130)

This 'long, hard look' reveals the exploitation, inequality and injustice dished out by mainstream tourism and profit-motivated tourist operators. The strategic counter-thinking to market forces highlighted by McLaren informs and shapes how we discuss the possibilities for slow ecotourist activism and change in local ecotourism through the following case study.

Slow Ecotourism: Peru

Visitors have a daunting array of things to do and places to see. It seems almost unthinkable not to visit Machu Pichu but there is so much else to do that some travellers miss this stunning archaeological highlight. Wildlife watchers might spend their entire visit ensconced in a rain forest lodge looking for toucans, turtles, and tapirs. Students can take courses in Spanish or Quechua or try their hands at some of the Indigenous crafts such as weaving. Outdoor enthusiasts can scale the highest peaks and surf the wildest beaches, visitors of all ages can learn to dance a sensual salsa to accompany the fabulous gastronomy that is slowly emerging onto the world culinary scene. (Rachowiecki, 2009: 8)

This wonderfully enthusiastic travel writing for a *National Geographic* guide to Peru evokes a country of tourist riches. As the front cover claims, Peru can offer 'off-the-beaten path excursions ... authentic experiences' that belie the fact that it is one of the world's poorest countries. Riches indeed are offered in the midst of poverty and continued political unrest, leaving the tourist reader wondering if such travel writing is not a smokescreen for smoothing over any hesitation to travel to such countries given the potential difficulties that might be faced. Such juxtaposition raises the ethical dilemmas and contradictions that are commonly associated with Western tourists travelling to developing countries such as Peru, where poverty, inequality and injustice are widespread.

Peru is a country facing many challenges in terms of poverty, political instability and lack of local economic and social infrastructure to cope with an increased ecotourism industry. Nonetheless, it is possible to imagine the provision of tourist services in local Peruvian communities that operate in slow, just and sustainable ways. This implies that the current practices create more inequalities rather than redistribute income in the community. Bauer (2008: 283) found that through 'personal communication in 2006 with a Peruvian village, it was uncovered the dismay of local people at the use of sacred knowledge and rituals deeply embedded in local culture on foreigners who lack the mental framework to respect Indigenous concepts'. In looking at mechanisms to reduce these occurrences, one avenue might be education, which can play an important role in the alleviation of misunderstandings between the host community and the tourist. Gulinck *et al.* (2001: 7) found that along with its potential to alleviate problems, education may also raise the quality of the experience for the tourist at the spiritual level 'and help them develop more of an awareness in relation to conservation and the protection of local cultures'.

In terms of tourism, Peru provides an enviable combination of picturesque countryside, World Heritage listed sites, long, sweeping coastline, glaciers and snow covered mountain peaks that reach over 6000 metres, extensive national parks and large tracts of jungle situated in the Amazon basin. These sites provide opportunities for deep experiences of ecological, cultural and archaeological tourism where trekking is the main form of transport. For example, if tourists wish to access the ancient city ruins of Machu Picchu, they need to trek on foot for three to four days on the Inca Trail from Cusco or use the PeruRail service. For some, the slow experience of trekking the Inca Trail embodies the principles of slow ecotourism and eco-tour operators. As is the case in tourist sites of other developing countries such as India and South Africa, walking around the cities and heritage sites gives a more considered and meaningful interaction with local mores, customs and culture. Unfortunately there is now a problem with the increasing numbers of tourists that visit these sites, indicating the limits to carrying capacity without appropriate development for slow ecotourism.

Ohl-Schacherer *et al.* (2008) discuss an Indigenous co-run ecotourism venture known as The Lodge in terms of sustainable practices such as profit sharing and employment opportunities for the Indigenous Matsiguenka community in the Manu National Park (that takes in part of the Amazon). The co-management of The Lodge between national parks and the Indigenous community has transformed the local economy so that cash benefits, through salaries and the sale of craft, and non-cash benefits, in the form of greater community organisation and integration into national Peruvian society, have resulted. Ohl-Schacherer *et al.*'s (2008: 14) broad evaluation of this ecotourism development remain positive in terms of social and economic benefits to the community, despite their reservations about cost effectiveness and building local infrastructure and building capacity.

Slow tourism at the point of consumption by the tourist and their host offers potential, but if viewed in terms of the market alone it can be seen as fraught with limited choices, for example, in terms of damage to the environment associated with travel to destinations. Jenner and Smith (2008) point to this problem in their green travel guide, suggesting that flying any distance by plane has a significant impact on the carbon footprint, so this is a limitation. They suggest that travelling by sailing boat or overland are much more sustainable and ethically justifiable options. However, this limits the range of travellers' opportunities.

Some Final Thoughts on Evaluating Slow Ecotourism

There is a need for benchmarks and indicators to accurately evaluate tourism impacts from a community perspective. Slow ecotourism requires an evaluation that represents integrity and accountability, one that maintains standards of socio-environmental impacts that are in line with a communities expressed values and sustainable parameters. More specifically, in relation to poverty alleviation, Manyara and Jones (2007) argue that assessment of community-based tourism initiatives should aim to measure: (1) the increase in direct income to households; (2) improvement in community services such as education (measured by increased literacy and numeracy levels), health services, clean water, appropriate housing, roads, transport and communication; and (3) the development of sustainable and diversified lifestyles. On this final point slow tourism can act as a platform that stimulates the creation of both tourism and non-tourism related small and medium size enterprises. Slow tourism provides the opportunity to re-evaluate and encompass social justice strategies and outcomes for local communities as it allows time to change, adjust and evolve for the tourism operator, community member and the environment.

We agree with anthropologists who argue that cultural lenses are necessary to understand the impact of tourism (amongst other market-based industries such as mining) on conservation and the environment in local host communities. This cultural approach requires several vantage points, including understanding conflicts rooted in culture, incorporating local cultural resources, and going beyond research empiricism in understanding Indigenous ecological knowledge. This can involve the use of ethnographic and culturally appropriate research methods, and 'hearing local voices speak', which includes outside evaluators who are not trying to speak for local communities (Peterson et al., 2010: 7–10) as well as including mechanisms for communities to self evaluate (see Wearing & McLean, 1997).

Lastly, the significance of social justice to slow ecotourism is reinforced by our analysis of commodification and tourism in general. Unless communities build tourist infrastructure for the purpose of creating more just and equal societies, a good deal of the social and economic benefits of local ecotourism will be lost to these communities (Binns & Nel, 2002; Ohl-Schacherer et al., 2008). Much stronger governmental controls, regulation and intervention is required. Unregulated development and the lack of existing environmental standards mean that Indigenous and host communities in the developed North and the impoverished South can only hope to slide into marketised solutions to poverty, hardship or the deprived living conditions of many of these communities, especially in developing countries. The point of slow travel is to counteract the pace of global market forces and to provide local spaces for the tourist–host encounter to experience a more authentic and deeper space that also has demonstrable cost saving benefits in a decreased use of fossil fuels, material products and cheap invited labour. Ideally, social justice based in greater recognition and social equality for these communities will push against the global inequalities of modern capitalism and the free market economy (Fraser, 2009).

Conclusion

We have argued that there are both challenges and possibilities in setting up slow agendas for ecotourism. Such agendas, we suggest, may help decommodify and de-objectify the host communities' experiences of tourism and tourists. How are decommodifying spaces for ecotourism created through appropriate social development and socially-just principles, processes and outcomes? In terms of the principles of slow ecotourism, without strong socio-environmental standards there can be no explicit government and non-government regulation and accountability of eco-tourist providers. Social justice needs to be at the heart of any economic or social changes developed around local eco-tourist enterprises and markets (Kosoy & Corbera, 2010; Sen, 2009). In terms of development processes, tour operators and hosts can learn from each other to nurture their environments and implement practices of sustainable and slow local ecotourism.

References

Adam, B. (2004) *Time*. Cambridge: Polity Press.

Al-Oun, S. and Al-Homoud, M. (2008) The potential for developing community-based tourism among the Bedouins in the Badia of Jordan. *Journal of Heritage Tourism* 3 (1), 36–54.

Aramberri, J. (2001) The host should get lost: Paradigms in the tourism theory. *Annals of Tourism Research* 28 (3), 738–761.

Bauer, I. (2008) The health impact of ecotourism on local indigenous populations in resource-poor countries. *Travel Medicine and Infectious Disease* 6, 276–291.

Beeton, S. (2006) *Community Development through Tourism*. Melbourne: Landlinks Press.

Binns, T. and Nel, E. (2002) Tourism as local development strategy in South Africa. *The Geographical Journal* 169 (3), 235–247.

Boulanger, P-M. (2009) Basic income and sustainable consumption strategies. *Basic Income Studies* 4 (2), 1–11.

Chan, Y.W. (2006) Coming of age of the Chinese tourists: The emergence of non-Western tourism and host-guest interactions in Vietnam's border tourism. *Tourist Studies* 6 (3), 187–213.

Cole, S. (2007) Implementing and evaluating a code of conduct for visitors. *Tourism Management* 28, 443–451.

Cunningham, P. (2006) Social valuing for Ogasawara as a place and space among ethnic host. *Tourism Management* 27 (3), 505–516.

Dickinson, J.E. and Lumsdon, L. (2010) *Slow Travel and Tourism*. Washington, DC: Earthscan.

Fennell, D.A. (2003) *Ecotourism: An Introduction* (2nd edn). London: Routledge.

Fraser, N. (2009) *Scales of Justice: Reimagining Political Space in a Globalizing World*. Columbia University Press.

Goodin, R.L., Rice, J.M., Parpo, A. and Eriksson, L. (2008) *Discretionary Time: A New Measure of Freedom*. Cambridge: Cambridge University Press.

Gulinck, H., Vyverman, N., Van Bouchout, K. and Gobin, A. (2001) Landscape as framework for integrating local subsistence and ecotourism: A case study in Zimbabwe. *Landscape and Urban Planning* 53, 173–182.

Hicks, A. and Esping-Andersen, G. (2005) Comparative and historical studies of social policy and the welfare state. In T. Janoski, R.R. Alford, A.M. Hicks and M.A. Schwartz (eds) *The Handbook of Political Sociology* (pp. 509–525). Cambridge: Cambridge University Press.

Hochschild, A. (2003) *The Commercialisation of Intimate Life*. Berkley: University of California Press.

Jenner, P. and Smith, C. (2008) *The Green Travel Guide: Your Passport to Responsible, Guilt Free Travel*. Richmond: Crimson Publishing.

Kosoy, N. and Corbera, E. (2010) Payment for ecosystem services as commodity fetishism. *Ecological Economics* 69, 1228–1236.

Liu, Z. (2003) Sustainable tourism development: A critique of sustainable tourism. *Journal of Sustainable Tourism* 11, 459–475.

Lyons, K.D. and Wearing, S. (eds) (2008) *Journeys of Discovery in Volunteer Tourism: International Case Study Perspectives*. Wallingford: CAB International.

MacEwan, A. (1999) *Neoliberalism or Democracy? Economic Strategy, Markets, and Alternatives for the 21st Century*. London: Zed Books.

Mandel, E. and Novack, G. (1970) *The Marxist Theory of Alienation*. New York: Pathfinder.

Manyara, G. and Jones, E. (2007) Community-based tourism enterprises development in Kenya: An exploration of their potential as avenues of poverty reduction. *Journal of Sustainable Tourism* 15 (6), 628–644.

Mbaiwa, J.E. (2004) The socio-cultural impacts of tourism development in the Okavango delta, Botswana. *Journal of Tourism and Cultural Change* 2 (3), 163–184.

McLaren, D. (1998) *Rethinking Tourism and Ecotravel: The Paving of Paradise and What You Can Do to Stop It*. West Hartford, CT: Kumarian Press.

Meethan, K. (2001) *Tourism in Global Society: Place, Culture, Consumption*. Basingstoke: Palgrave.

Milne, S. and Ateljevic, I. (2001) Tourism, economic development and the global–local nexus: Theory embracing complexity. *Tourism Geographies* 3 (4), 369–393.

Ohl-Schacherer, J., Mannigel, E., Kirby, C., Shepard, G.H. and Douglas, W.Y. (2008) Indigenous ecotourism in the Amazon: A case study of 'Casa Matsiguenka' in Manu National Park, Peru. *Environmental Conservation* 35 (1), 14–25.

Parkins, W. and Craig, G. (2006) *Slow Living*. Sydney: University of New South Wales Press.

Peterson, R.B., Russle, D., West, P. and Brosius, J.P. (2010) Seeing and doing conservation through cultural lenses. *Environmental Management* 45 (5), 5–18.

Rachowiecki, R. (2009) *Peru: National Geographic Traveler*. Washington, DC: National Geographic Society.

Sen, A. (2009) *The Tacit Dimension*. London: University of Chicago Press.

Simmel, G. (1990) *The Philosophy of Money* (T. Bottomore and D. Frisby, trans.). London: Routledge (original work published 1907).

Simmel, G. (1991) Money in modern culture. *Theory, Culture & Society* 8 (3), 17–31.

Sherlock, K. (2001) Revisiting the concept of hosts and guests. *Tourist Studies* 1 (3), 271–295.

Sinclair, M.T. (1998) Tourism and economic development: A survey. *Journal of Development Studies* 34 (5), 1–51.

Taylor, J.P. (2001) Authenticity and sincerity in tourism. *Annals of Tourism Research* 28 (1), 7–26.

van der Duim, R., Peters, K. and Wearing, S. (2005) Planning host and guest interactions: Moving beyond the empty meeting ground in African encounters. *Current Issues in Tourism* 8 (4), 286–305.

Wearing, S. and McDonald, M. (2002) The development of community based tourism: Re-thinking the relationship between tour operators and development agents as intermediaries in rural and isolated area communities. *Journal of Sustainable Tourism* 10 (3), 191–206.

Wearing, S. and McLean, J. (1997) *Developing Ecotourism: Community Based Approach*. Newport, Victoria: HM Leisure Planning Pty Ltd.

Wearing, S. and Neil, J. (2009) *Ecotourism: Impacts, Potentials and Possibilities* (2nd edn). Oxford: Butterworth-Heinemann.

Wearing, S. and Wearing, M. (1999) Decommodifying ecotourism: Rethinking global-local interactions with host communities. *Leisure and Society* (*Loisire et Societe*) 22 (1), 39–70.

Williams, A.M. and Shaw, G. (eds) (1999) *Tourism and Economic Development: European Experience* (3rd edn). New York: Wiley.

World Wildlife Fund (2010) WWF – Who we are – Online document: http://www.worldwildlife.org/who/Vision/index.html

World Tourism Organisation (2002) *Ecotourism Market Reports – Set of Seven Reports*. Madrid: UNWTO Publications.

Zaccai, E. (ed.) (2007) *Sustainable Consumption: Ecology and Fair Trade*. London: Routledge.

Part 2
Slow Food and Sustainable Tourism

5 The Contradictions and Paradoxes of Slow Food: Environmental Change, Sustainability and the Conservation of Taste

C. Michael Hall

The Slow Food movement is arguably one of the focal points of the broader interest in slow consumption and lifestyles. This approach suggests that for sustainability to be achieved technological improvements alone will not suffice and instead there is a need to slow the rate at which raw materials are transformed into products and services and eventually discarded as 'waste' (Cooper, 2005). Although slow consumption was initially focussed on more labour-intensive production practices that allowed for the retention and creation of eco-efficient, decentralised and cultural and natural resource-preserving jobs and products (Ax, 2001), the concept has now been considerably expanded in scope (Cooper, 2010; Hall, 2009). The purpose of this chapter is to examine the eco-gastronomy of the Slow Food movement which is often portrayed as a significant contribution to food security and the maintenance of heritage varieties, foodways and lifestyles. In doing so, it seeks to understand how the Slow Food movement might be able to contribute to the development of more sustainable forms of tourism and hospitality and more ecologically and socially responsible forms of consumption.

Consumption

Consumption is a pervasive element of social, economic and political organisation in the modern world. However, consumption has become increasingly problematic in light of the potential ecological harm of overconsumption of renewable natural resources and the cultural and economic effects of particular consumptive practices. These, in turn, are associated with the nature of contemporary capitalism and globalisation in which consumption becomes a goal for its own sake. This situation – of the link between ecological degradation, consumption and the prevailing economic and political institutions in which consumers are unaware of the use of natural resources in the production of goods and services – has been termed 'hyperconsumption', where 'the sign value, or image, eclipses the commodity referent and simultaneously negates the ecological referent of the commodity as a product of nature' (Kilbourne et al., 1997: 8). For Dolan (2002) the explanation for such hyperconsumption is located within the dominant social paradigm, which includes ideologies of progress and rationality, and in which the ecological referent of products has been lost. One reaction to both the environmental state that has developed as a result of consumptive practices as well as the behaviours that underlie such practices has been an emergence of interest in the notion of sustainable consumption.

Sustainable consumer practices can range from the act of purchasing fair trade coffee (which in itself can be a form of sign value) through to the growth of more systematic forms of anti-consumerism that have developed to counter the excesses of hyperconsumption and work intensified lifestyles that occupy time. A widely used definition of sustainable consumption is that of the 1994 Oslo Symposium on Sustainable Consumption: 'the use of goods and services that respond to basic needs and bring a better quality of life, while minimising the use of natural resources, toxic materials and emissions of waste and pollutants over the life cycle, so as not to jeopardise the needs of future generations' (Norwegian Ministry of the Environment, 1994: Sec. 1.2). More recently, Cooper (2005) has utilised the Organisation for Economic Cooperation and Development (OECD, 2002) definition 'the consumption of goods and services that meet basic needs and quality of life without jeopardizing the needs of future generations'. As Cooper (2005: 52) noted, both definitions could be interpreted in many different ways, in the same fashion as the concept of sustainable development of which they are an offshoot, 'but there is a general consensus that for industrialized countries, at least, it implies a reduction in the throughput of resources'.

This is not to suggest of course that the notion of sustainable consumption is completely new. Arguably there has been a long history of consumer inspired economic and political responses to consumption, ranging from the home economics movement that emerged in the USA in the late 19th century (Baldwin, 1991; Weigley, 1974), through to the development of national and international demands for increased consumer rights (Furlough & Strikwerda, 1999). Moreover, there is also a wealth of research interest in the implications of consumer action with respect to environmental, political and social causes dating back to the 1960s and 1970s. For example, Webster (1975: 188) defined the socially conscious consumer as 'a consumer who takes into account the public consequences of his or her private consumption or who attempts to use his or her purchasing power to bring about social change'. Roberts (1993: 140) expanded the notion of conscious consumption to refer to the socially responsible consumer as 'one who purchases products and services perceived to have a positive (or less negative) influence on the environment or who patronizes businesses that attempt to effect related positive social change'.

Interest in individual responsible consumption and its connection to business practices can also be identified in the development of social and environmental marketing concepts (Andreasen, 2003). Although the emphasis here is arguably connected more with tapping consumer demand than with being genuinely sustainable (Peattie & Crane, 2005). For example, the connection to corporate socially responsible activities is illustrated in Kotler's (1991) societal marketing concept, which is doing business in a way that maintains or improves both the customer's and society's well-being (however defined). Mohr *et al.* (2001: 47) used this concept to define corporate social responsibility as 'a company's commitment to minimizing or eliminating any harmful effects and maximizing its long-run beneficial impact on society', and went on to define socially responsible consumer behaviour as 'a person basing his or her acquisition, usage, and disposition of products on a desire to minimize or eliminate any harmful effects and maximize the long-run beneficial impact on society' (Mohr *et al.*, 2001: 47).

Such diverse perspectives on sustainable consumption have also been reflected in the tourism literature, although there is arguably much less of an emphasis on consumer activism compared with a more general notion of consumer values and their relationship to different types of consumptive practices (Hall & Brown, 2006). This is a situation which is perhaps even more pronounced in the hospitality literature (Hall, 2010a). Nevertheless, among the broad discourses of sustainability in tourism there is a gradual recognition of the potential significance of consumer movements and activism as part of the social construction of a more sustainable future for tourism (Caletrio, 2011; Gössling *et al.*, 2009; Hall, 2009, 2011; Zapata *et al.*, 2011).

Hall (2009, 2010b) has argued that within tourism there are primarily two ways to encourage a reduction of the environmental footprint of tourism. The first is the efficiency approach, which seeks to reduce the rate of consumption by using materials and resources more productively. Eco-efficiency stresses the technological link between value creation in economic activities and environmental quality. This approach places more focus on recycling, using energy more efficiently, eco-innovation, and reducing emissions, but otherwise operating in a 'business as usual' manner. Examples of this include the development of more efficient aircraft and infrastructure that reduces the throughput of energy and therefore emissions per consumer without reducing the overall level of demand (Gössling *et al.*, 2010).

The second approach has been referred to as 'slow consumption' (Cooper, 2005), also closely related to the concepts of *'décroissance'* or *'degrowth'*

(Bourdeau & Berthelot, 2008; Flipo & Schneider, 2008; Hall, 2009), and includes consumer activism as well as industry and public policy initiatives. These include (Hall, 2007):

- the development of environmental standards at the regional, national and international scales, e.g. such as the Nordic Swan label (e.g. Bohdanowicz, 2009) (also utilised under the efficiency approach);
- relocalisation schemes, such as farmers markets and 'local diets', that reinforce the potential economic, social and environmental benefits of purchasing, consuming and producing locally (e.g. Hall & Sharples, 2008);
- ethical consumption, through ethical and responsible tourism (e.g. Hall & Brown, 2006); and
- the so-called 'new politics of consumption' such as anti-consumerism and culture jamming (e.g. Cohen et al., 2005).

The latter approach in particular aims to position tourism as part of a circular economy rather than a linear one, so that inputs of virgin raw material and energy and outputs in the form of emissions and waste requiring disposal are reduced (Cooper, 2005; Hall, 2009, 2010b; Jackson, 2005). At the forefront of the development of slow tourism, and what is often regarded as an exemplar (Dickinson & Lumsdon, 2010; Hall, 2006), is the 'Slow Food movement'. Established in Italy in 1989 by Carlo Petrini in response to the establishment of the first McDonald's restaurant in Rome, the Slow Food movement now has international significance with respect to the broader understanding of the sustainability of foodways and food-related biodiversity as well as the broader discourses of slow consumption (Cooper, 2005). The next section provides an overview of the goals, strategies and activities of the Slow Food movement before examining its significance for tourism and hospitality.

Slow Food

Slow Food is a non-profit, member-supported association formed in Italy in 1989 and now claims supporters in 150 countries, over 100,000 members, with 1300 *convivia* (local chapters), as well as a network of 2000 food communities who practice small-scale and sustainable production of quality foods. Slow Food originally located its philosophical origins in the 17th-century writings of Francesco Angelita, who considered slowness a virtue and, believing that all creatures bore messages from God, wrote a book about snails. Slow Food thus adopted a snail as its symbol (Slow Food, 2002, cited in Hall, 2007). At the time of writing, the philosophy of Slow Food as stated on their international website reflected why the organisation is often perceived as part of the anti-globalisation movement.

Slow Food stands at the crossroads of ecology and gastronomy, ethics and pleasure. It opposes the standardization of taste and culture, and the unrestrained power of the food industry multinationals and industrial agriculture. We believe that everyone has a fundamental right to the pleasure of good food and consequently the responsibility to protect the heritage of food, tradition and culture that make this pleasure possible. Our association believes in the concept of neo-gastronomy – recognition of the strong connections between plate, planet, people and culture.

Our Vision
We envision a world in which all people can access and enjoy food that is good for them, good for those who grow it and good for the planet.

Our Mission
Slow Food is an international grassroots membership organization promoting good, clean and fair food for all.

Good, Clean and Fair
Slow Food's approach to agriculture, food production and gastronomy is based on a concept of food quality defined by three interconnected principles:

GOOD a fresh and flavorsome seasonal diet that satisfies the senses and is part of our local culture;
CLEAN food production and consumption that does not harm the environment, animal welfare or our health;
FAIR accessible prices for consumers and fair conditions and pay for small-scale producers. (Slow Food, 2010)

The notion of clean in Slow Food philosophy usually also means that foods have been cultivated in an environmentally sustainable fashion, are local and are not genetically modified (Earth Markets, 2010). Slow Food also promote the importance of consumer choice to bring about change, and promote the idea of members and supporters as co-producers 'an eater who is informed about where and how their food is produced and actively supports local producers, therefore becoming part of the production process'. They also indicate support for local identity via the protection of 'traditional and sustainable quality foods, defending the biodiversity of cultivated and wild varieties as well as cultivation and processing methods', as well as the importance of network development 'Slow Food defends biodiversity in our food supply, promotes food and taste education and connects sustainable producers to co-producers through events and building networks' (Slow Food, 2010).

In order to give effect to its philosophy, Slow Food has five main strategies that are outlined below.

(1) Terra Madre network

Terra Madre (Mother Earth) is a network of individuals, organisations and businesses that operates at various scales. The aim of the network is to 'give a voice and visibility to the small-scale farmers, breeders, fishers and food artisans around the world whose approach to food production protects the environment and communities' (Slow Food, 2010). The Terra Madre website states that the food community network 'brings together those players in the food chain who together support sustainable agriculture, fishing, and breeding with the goal of preserving taste and biodiversity' (Terra Madre, 2010). The network therefore represents a conduit for the exchange of sustainable food related knowledge as well as opportunities to develop new short food chains between agricultural producers, restaurants and consumers.

(2) Biodiversity

A Slow Food Foundation for Biodiversity was founded in 2003 to develop projects and new economic models to support the Terra Madre food communities and networks. The Foundation's projects are focused on protecting food biodiversity and traditions, and promoting sustainable agriculture. Three main projects exist. First, the Presidia, which are sustainable food production initiatives that build the capacity of a group of producers in order to improve production techniques, develop production protocols and find local and international markets. As of November 2010, more than 300 Presidia had been created around the world, involving over 10,000 small-scale sustainable farmers. Each Presidium supports a quality product at risk of extinction; uses traditional processing and/or agricultural methods; and safeguards native breeds and local plant varieties (Slow Food, 2010). Second, the Ark of Taste, which was established in 1996, and which aims to (re)discover, catalogue and promote foods which are at risk of being lost through either biodiversity loss or industrial standardisation, but that have productive and commercial potential and are closely associated with specific communities and cultures. As of 2010 more than 900 foods from over 50 countries had been listed (Slow Food Foundation, 2010). The third, and most recent, project is the Earth Markets, which are farmers' markets run according to certain Slow Food principles and which are overseen by a community management group (Earth Markets, 2010). According to the Earth Markets website, such markets 'strengthen local food networks' and provide:

- Quality food you can TRUST, bought directly from the producers.
- Fair prices for both consumers and producers that foster LOCAL ECONOMIES.

- Access to good, clean and fair food from the local area to reduce food miles and SHORTEN the FOOD CHAIN.
- Consumers become COPRODUCERS, learning from producers and EDUCATIONAL activities. (Earth Markets, 2010)

As of November 2010 there were 15 Earth Markets in five countries.

(3) Food and taste education

Slow Food engages in 'good, clean and fair' food education with respect to both tasting food and sustainability issues in the food chain. Education includes working with schools, local producers, conferences and community education.

(4) Connecting producers and consumers

Slow Food uses the term 'co-producer' to refer to the way in which an individual 'goes beyond the passive role of a consumer to take an active interest in who producers our food, their methods, the problems they face and the impact on the world around us' (Slow Food, 2010). To encourage connections between producers and consumers Slow Food focuses on the development of national and international events to showcase producers' food and create opportunities for consumers to meet producers. International events include Salone Del Gusto, Cheese, Slow Fish and Terra Madre, all held in Italy. Salone Del Gusto attracts over 150,000 visitors to Turin, Italy every two years, while Cheese, held in Bra, attracts over 100,000 visitors in the course of the four-day event (Slow Food, 2010). National and major regional events are held in Germany, Ireland, Spain, France and Japan, while there are 1300 Slow Food *convivia* groups around the world that also provide opportunities for developing producer and consumer networks.

(5) Campaigns

As of the end of 2010 Slow Food was engaged in six different campaigns:

Slow Canteens which aims to introduce good, clean and fair food to public canteens in schools, hospitals and workplaces in order promote quality everyday food and education. This has led to projects such as Slow Food in the Canteen (2010), a European network of schools established in 2009 that works to improve student meals and increase awareness of food issues and that is operating in 12 schools in 10 European countries. Slow Food USA have run a Time for Lunch campaign aimed at improving school food and lobbied the American Congress on child health issues in its revision of the Child Nutrition Act and providing funding for federal child nutrition programmes, including the National School Lunch programme. Slow Food France and Italy have also launched campaigns to improve canteen food (Slow Food, 2010).

Food Sovereignty refers to the promotion of the 'right to choose what makes up our daily diet for all people, including the knowledge and freedom to choose what to grow, and how food is treated and distributed' (Slow Food, 2010). This has been addressed by the production of various manifestos as well as working with a coalition of organisations on the European Food Declaration (2010), which aims to develop a fairer, more inclusive and sustainable food system in Europe and that challenges the current Common Agriculture Policy (CAP) and the European Commission's current plans for a renewed CAP in 2013. The signatories to the European Food Declaration (2010) believe 'a new Common Food and Agriculture Policy should guarantee and protect citizens' space in the EU and candidate countries and their ability and right to define their own models of production, distribution and consumption' following a set of principles outlined in Figure 5.1. According to Slow Food (2010) many of Slow Food's food security, Presidia and education projects work towards the same goals.

Land Grabbing is the Slow Food Foundation for Biodiversity campaign in conjunction with a coalition of organisations to protest against the increase in land grabbing – transferring rights over agricultural land in developing countries to foreign investors – which can pose a serious threat to food sovereignty and community rights, and to denounce its support by the World Bank (Slow Food, 2010).

Next Generation is a campaign to educate and motivate younger people 'to take an active interest and role in their food future, and to find opportunities for knowledge to be passed down by elders in communities around the world to the next generation' (Slow Food, 2010). Components include various school garden projects; Slow Food on Campus which operates in Canada, the UK and USA; a Youth Food Movement; Terra Madre Youth; Terra Madre Young Europeans; and a Grandmothers Day that was held in Ireland to draw attention to the preservation of inter-generational food knowledge.

GMOs. Slow Food is extremely active globally in banning the commercial planting of genetically modified crops (GMOs) and promoting GMO-free food and animal feed. Opposition ranges over a number of factors including application of the precautionary principle, economic dependence of food producers on seed suppliers, need for full labelling, and cross contamination.

Raw Milk is one of the oldest Slow Food campaigns. Launched in 2001 the campaign aims to protect the rights of cheese makers to produce raw milk cheese where hygiene laws may forbid raw cheese production or make it extremely difficult.

The Contribution of Slow Food to Slow Tourism and Hospitality

Slow Food has been integral to the promotion of the slow concept. In fact, it probably would not exist without it, in the popular imagination at

(1) Considers food as a universal human right, not merely a commodity.

(2) Gives priority to growing food and feed for Europe and changes international trade in agricultural products according to principles of equity, social justice and ecological sustainability. The CAP should not harm other countries' food and agriculture systems.

(3) Promotes healthy eating patterns, moving towards plant-based diets and towards a reduced consumption of meat, energy-dense and highly processed foods, and saturated fats, while respecting the regional cultural dietary habits and traditions.

(4) Gives priority to maintaining an agriculture all over Europe that involves numerous farmers producing food and caring for the countryside. That is not achievable without fair and secure farm prices, which should allow a fair income for farmers and agricultural workers, and fair prices for consumers.

(5) Ensures fair, non-discriminatory conditions for farmers and agricultural workers in Central and Eastern Europe, and promotes a fair and equitable access to land.

(6) Respects the local and global environment, protects the finite resources of soil and water, increases biodiversity and respects animal welfare.

(7) Guarantees that agriculture and food production remain free from GMOs and fosters farmers' seeds and the diversity of domestic livestock species, building on local knowledge.

(8) Stops promoting the use and the production of industrial agrofuels and gives priority to the reduction of transport in general.

(9) Ensures transparency along the food chain so that citizens know how their food is produced, where it comes from, what it contains and what is included in the price paid by consumers.

(10) Reduces the concentration of power in the agricultural, food processing and retail sectors and their influence on what is produced and consumed, and promotes food systems that shorten the distance between farmers and consumers.

(11) Encourages the production and consumption of local, seasonal, high quality products reconnecting citizens with their food and food producers.

(12) Devotes resources to teaching children the skills and knowledge required to produce, prepare, and enjoy healthy, nutritious food.

Figure 5.1 European Food Declaration set of principles for a new EU Common Food and Agricultural Policy
Source: European Food Declaration (2010).

least. Slow Food followers argue for the importance of preserving the sensual pleasures and qualities of everyday life from the relentless pursuit of speed and the retail notion of convenience by slowing consumption down and celebrating and supporting traditional lifestyles and foodways (Cohen, 2006; Jones *et al.*, 2003; Kummer, 2002). Osborne (2001: 100) described Slow Food as the 'gastronomic version of Greenpeace: a defiant determination to preserve unprocessed, time-intensive food from being wiped off the culinary map'. However, such a comparison while perhaps true with respect to public profile is certainly not valid with respect to tactics.

Although Slow Food does share some of the rhetoric of anti-globalisation activism it does not embrace the political dimension of a Greenpeace environmental campaign or Adbusters anti-advertising (Andrews, 2008; Miele & Murdoch, 2002). It certainly does not embrace the overt direct action of an organisation such as Sea Shepherd. Instead, Slow Food is focussed on more of a lifestyle reinvention approach which is more inward-looking and focussed on the transformation of personal values and practices (Cohen, 2006). This is clearly evidenced by the emphasis of Slow Food on education and knowledge transfer via network development and events and not on political protest and confrontation. Slow Food members are much more likely to be engaged in letter writing and lobbying than marching. Although this may be relevant to the everyday politics of food consumption and may help change the market logic of food, at least in part, with respect to the valorisation of local, organic and heritage foods, it does not yet represent a more fundamental change to food systems. Instead, the more significant change to food systems, like the emphasis of Slow Food itself, tends to be highly localised. Indeed, the extent to which Slow Food has a broader affect on consumption practices remains little explored. Nevison's (2008) survey of 309 Slow Food members from 13 countries did find that some members had expanded the concept of 'slow' beyond food, with 54% trying to incorporate 'slow' principles into transportation, 47.2% into vacation and travel, and 44% into shelter/housing, with a few respondents also including energy in their 'slow' agenda. For some, Slow Food may therefore be a starting point into a wider approach to slow consumption. Nevertheless, as Gaytán (2004: 97) noted, American members of Slow Food 'manage multiple identities in an attempt to resist and mobilize against the negative consequences of industrialization. ... Despite the creation of an innovative site of resistance, Slow Food members construct a limited notion of the local that excludes working-class and urban cultural expressions'.

The contribution of Slow Food to the development of a notion of Slow Tourism should therefore be considered at a number of different levels. National and international Slow Food events are themselves tourist events, as the number of attendees noted above highlights. Therefore, it is perhaps not surprising that cities are interested in hosting and supporting such meetings. For example, the eighth Salone del Gusto in October 2010, Slow Food's biannual food festival held in Turin, attracted 200,000 visitors of which 30% were estimated to have come from outside of Italy. Moreover, tourism businesses, such as holiday package companies, hotels and hotel chains, car rental companies and destinations, are interested in associating themselves with Slow Food by advertising in their publications and on their websites. Of course, such a situation has raised questions about the extent to which Slow Food actually represents fundamental change and whether it is just 'elitist and effete, too expensive for ordinary people, just the latest trend among foodies and gourmands' (Schlosser, 2002: 11), as some critics have claimed (Singh, 2005; White, 2008).

Nevertheless, the Slow Food movement has been successful in raising the profile of a number of issues, particularly the significance of local food and fair trade purchasing for restaurants and hotels as well as consumers in general, and the transformation of local farmers' markets and direct food purchase from growers into a leisure commodity. These two contributions are not the result of Slow Food alone, however, and are instead part of a wider consumer movement associated with ethical consumption and the relocalisation of consumption (Cohen, 2006; Hall, 2010a; Parkins & Craig, 2006), but with specific focus on local food systems or foodsheds (Hall & Sharples, 2008; Kimura & Nishiyama, 2008) and food miles and the carbon footprint of food (Gössling, 2011; Hall & Sharples, 2008).

A local food system refers to deliberately formed systems that are characterised by a close producer–consumer relationship within a designated place or local area. Local food systems support long-term connections; meet economic, social, health and environmental needs; link producers and markets via locally focussed infrastructure; promote environmental health; and provide competitive advantage to local food businesses (Buck *et al.*, 2007; Food System Economic Partnership, 2006). A similar concept is that of the 'foodshed', which is associated with the bio-regionally-oriented concept of the watershed (Feagan & Krug, 2004; Kloppenburg *et al.*, 1996). Direct marketing via farmers' markets and food festivals, along with other forms of tourism, are recognised as integral components of a foodshed (Hall & Sharples, 2008). According to Feagan and Krug (2004) in order for a local foodshed to be established, several things need to happen:

* producers and consumers must be brought closer together to shorten food chains and to build 'community' and foster sustainability;
* there must be public awareness of the nature of the 'costs' associated with the industrial food system so that local consumers and producers will rethink their food production and purchasing decisions; and
* the means – mechanisms, places and opportunities – for meeting objectives must be made available.

According to Anderson and Cook (2000: 237):

> The major advantage of localizing food systems, underlying all other advantages, is that this process reworks power and knowledge relationships in food supply systems that have become distorted by increasing distance (physical, social, and metaphorical) between producers and consumers. … [and] gives priority to local and environmental integrity before corporate profit-making.

However, despite the profile given to the relocalising of the food system and the Slow Food message, the adoption of many of its projects remains piecemeal (Singh, 2005). The experience of Slow Food with respect to the promotion of localised food consumption therefore reflects that of the broader local food movement in that many consumer and small producer-based initiatives become organised by governments and larger producer organisations. Indeed, while Slow Food has added some important resources and a conceptual framework and philosophy for the ethical consumption of food, like the *chisan-chisho* movement in Japan (Kimura & Nishiyama, 2008), there is a danger that its initiatives become refashioned as a producer movement by government and destination marketing and development bodies so as to capitalise on local food's marketing appeal. Slow Food, along with other similar organisations, has not to date been able to develop a full-fledged, citizen-based political mobilisation, or address the issue of marginality in the food system (Kimura & Nishiyama, 2008; Van der Meulen, 2008). Moreover, there are broader questions over whether material pleasure and the symbolic expression of identity through good taste and the consumption of food is compatible with a more politicised, socially conscious consumption ethos (Pietrykowski, 2004).

Nevertheless, the main argument to counter the criticism of marginality is that local food networks serve as an important signal and example to the mainstream, reflecting where society is going and where new opportunities for consumption and production lie (Van der Meulen, 2008). 'The actual practices do not represent a simple re-proposing of old traditional productions, rather they derive from a new reading of the internal and external environment based on the needs and characters of the modern consumer' (Nosi & Zanni, 2004: 789). Therefore, probably some of the largest gains are likely to be realised by actors outside the initial Slow Food networks (Van der Meulen, 2008). In the same vein, conspicuous consumption of local specialty food products by upper-class people will turn local foods into 'culture goods' (Bourdieu, 1984), making them more broadly desirable to other classes. Clearly, such a situation may well fulfil many of the Slow Food movement's goals (Petrini, 2001, 2007). Yet in trying to conceive of a Slow Food inspired practice of tourism many issues remain. Most significantly is the extent to which Slow Food actually represents a move towards a more sustainable form of travel consumption. Unfortunately, this is probably not the case, as the movement appears unaware of the potential contradictions between mobility and sustainability, as Petrini (2007: 241) writes:

> It is necessary to *move*, to meet people, to experience other territories and other tables. If we apply this conviction to the network it is vital to guarantee the circulation within it of people, from one side of the globe to the other, without distinction and without restriction. The *right to travel* becomes fundamental, a premise on which to base cultural growth and the self-nourishment of the network of gastronomes.

Moreover, even if one ignores the potential emissions of travel, there remains a significant issue in that in many cases the local food system still requires distant consumers to make artisan foods economic. In other words, there is a need to consider local food systems, as well as consumption, within the international economic and regulatory system in which they are framed. The Slow Food experience therefore suggests that much of the discussion with respect to slow tourism needs to be understood as an issue of critical consumption rather than sustainable consumption. That is, it is more an expression of public discomfort than a real reformulation of lifestyle (Cohen, 2006; Dickinson & Lumsdon, 2010), although clearly some members want to live a more sustainable life (Nevison, 2008).

More fundamental questions with respect to population growth, income (re)distribution and restrictions of consumption outside of the food domain are not yet asked within the Slow Food mantra. Yet they are probably the most difficult, and most important, questions of all as they imply that for the 'inclusive elite' of Slow Food – as Petrini prefers to call Slow Food members (Van der Meulen, 2008: 235) – there is a need to redistribute and restrict some of their consumptive activities, including the supposed right to travel. To put it simply, calls for more efficacious lifestyles among residents of the world's developed countries, including genuinely sustainable travel practices, will continue to be resisted by policy-makers because sustainable consumption runs counter to dominant tenets of neo-liberal economics and conventional political objectives (Cohen, 2006). Indeed, it is the systemic nature of Slow that most requires critique (Hall, 2010b), as well as its seemingly inevitable commodification as yet another marketing slogan for screwing the Earth.

References

Anderson, M.D. and Cook, J.T. (2000) Does food security require local food systems? In J.M. Harris (ed.) *Rethinking Sustainability: Power, Knowledge and Institutions* (pp. 228–248). Ann Arbor, MI: University of Michigan Press.

Andreasen, A.R. (2003) The life trajectory of social marketing. Some implications. *Marketing Theory* 3 (3), 293–303.

Andrews, G. (2008) *The Slow Food Story: Politics and Pleasure*. Montréal: McGill-Queen's University Press.

Ax, C. (2001) Slow consumption for sustainable jobs. In M. Charter and U. Tischner (eds) *Sustainable Solutions* (pp. 402–409). Sheffield: Greenleaf.

Baldwin, E.E. (1991) The home economics movement: A 'new' integrative paradigm. *Journal of Home Economics* 83 (4), 42–49.

Bohdanowicz, P. (2009) Theory and practice of environmental management and monitoring in hotel chains. In S. Gössling, C.M. Hall and D. Weaver (eds) *Sustainable Tourism Futures: Perspectives on Systems, Restructuring and Innovations* (pp. 102–130). New York: Routledge.

Bourdieu, P. (1984) *Distinction*. London: Routledge.

Bourdeau, P. and Berthelot, L. (2008) Tourisme et décroissance: de la critique à l'utopie? In F. Flipo and F. Schneider (eds) *Proceedings of the First International Conference on Economic De-growth for Ecological Sustainability and Social Equity, Paris, 18–19 April 2008* (pp. 78–85). Paris.

Buck, K., Kaminski, L.E., Stockmann, D.P. and Vail, A.J. (2007) *Investigating Opportunities to Strengthen the Local Food System in Southeastern Michigan, Executive Summary.* MI: University of Michigan, School of Natural Resources and Environment.

Caletrio, J. (2011) Simple living and tourism in times of 'austerity'. *Current Issues in Tourism.* DOI: 10.1080/13683500.2011.556246.

Cohen, M.J. (2006) Sustainable consumption research as democratic expertise. *Journal of Consumer Policy* 29 (1), 67–77.

Cohen, M.J., Comrov, A. and Hoffner, B. (2005) The new politics of consumption: Promoting sustainability in the American marketplace. *Sustainability: Science, Practice, and Policy* 1, 58–76.

Cooper, T. (2005) Slower consumption: Reflections on product life spans and the 'throw-away society'. *Journal of Industrial Ecology* 9, 51–67.

Cooper, T. (ed.) (2010) *Longer Lasting Products: Alternatives to the Throwaway Society.* Farnham: Gower.

Dickinson, J.E. and Lumsdon, L. (2010) *Slow Travel and Tourism.* London: Earthscan.

Dolan, P. (2002) The sustainability of 'sustainable consumption'. *Journal of Macromarketing* 22, 170–181.

Earth Markets (2010) Online document: http://www.earthmarkets.net/

European Food Declaration (2010) Online document: http://www.europeanfood declaration.org/

Feagan, R. and Krug, K. (2004) Towards a sustainable Niagara foodshed: Learning from experience. In *Leading Edge 2004 The Working Biosphere*, 3–5 March. Georgetown: Niagara Escarpment Commission.

Flipo, F. and Schneider, F. (eds) (2008) *Proceedings of the First International Conference on Economic De-growth for Ecological Sustainability and Social Equity, Paris, 18–19 April 2008.* Paris.

Food System Economic Partnership (2006) *Alternative Regional Food System Models: Successes and Lessons Learned: A Preliminary Literature Review.* MI: Food System Economic Partnership.

Furlough, E. and Strikwerda, C. (eds) (1999) *Consumers Against Capitalism? Consumer Cooperation Europe, North America, and Japan, 1840–1990.* Lanham, MD: Rowman and Littlefield.

Gaytán, M.S. (2004) Globalizing resistance: Slow Food and new local imaginaries. *Food, Culture and Society: An International Journal of Multidisciplinary Research* 7 (2), 97–116.

Gössling, S. (2011) *Carbon Management in Tourism: Mitigating the Impacts on Climate Change.* London: Routledge.

Gössling, S., Hall, C.M., Peeters, P. and Scott, D. (2010) The future of tourism: A climate change mitigation perspective. *Tourism Recreation Research* 35 (2), 119–130.

Gössling, S., Hall, C.M. and Weaver, D. (eds) (2009) *Sustainable Tourism Futures: Perspectives on Systems, Restructuring and Innovations.* New York: Routledge.

Hall, C.M. (2006) Culinary tourism and regional development: From slow food to slow tourism? *Tourism Review International* 9 (4), 303–306.

Hall, C.M. (2007) The possibilities of slow tourism: Can the slow movement help develop sustainable forms of tourism consumption? Paper presented at 'Achieving Sustainable Tourism', 11–14 September, Helsingborg, Sweden.

Hall, C.M. (2009) Degrowing tourism: Décroissance, sustainable consumption and steady-state tourism. *Anatolia: An International Journal of Tourism and Hospitality Research* 20 (1), 46–61.

Hall, C.M. (2010a) Blending fair trade coffee and hospitality. In L. Jolliffe (ed.) *Coffee Culture, Destinations and Tourism* (pp. 159–171). Bristol: Channel View Publications.

Hall, C.M. (2010b) Changing paradigms and global change: From sustainable to steady-state tourism. *Tourism Recreation Research* 35 (2), 131–145.

Hall, C.M. (2011) Policy learning and policy failure in sustainable tourism governance: From first and second to third order change? *Journal of Sustainable Tourism* 19 (4–5), 649–671. DOI: 10.1080/09669582.2011.555555.

Hall, C.M. and Sharples, L. (2008) Food events and the local food system: Marketing, management and planning issues. In C.M. Hall and L. Sharples (eds) *Food and Wine Festivals and Events Around the World: Development, Management and Markets* (pp. 23–46). Oxford: Butterworth Heinemann.

Hall, D. and Brown, F. (2006) *Tourism and Welfare: Ethics, Responsibility and Sustainable Well-Being*. Wallingford: CABI.

Jackson, T. (2005) Live better by consuming less? Is there a 'double dividend' in sustainable consumption. *Journal of Industrial Ecology* 9 (1–2), 19–36.

Jones, P., Shears, P., Hillier, D., Comfort, D. and Lowell, J. (2003) A return to traditional values? A case study of slow food. *British Food Journal* 105 (4–5), 297–304.

Kilbourne, W., McDonagh, P. and Prothero, A. (1997) Sustainable consumption and the quality of life: A macromarketing challenge to the dominant social paradigm. *Journal of Macromarketing* 17, 4–24.

Kimura, A.H. and Nishiyama, M. (2008) The chisan-chisho movement: Japanese local food movement and its challenges. *Agriculture and Human Values* 25 (1), 49–64.

Kloppenburg, J., Hendrickson, J. and Stevenson, G. (1996) Coming into the foodshed. *Agriculture and Human Values* 13, 23–32.

Kotler, P. (1991) *Marketing Management: Analysis, Planning, Implementation, and Control* (7th edn). Englewood Cliffs, NJ: Prentice Hall.

Kummer, C. (2002) *The Pleasures of Slow Food: Celebrating Authentic Traditions, Flavours and Recipes*. New York: Chronicle Books.

Miele, M. and Murdoch, J. (2002) The practical aesthetics of traditional cuisines: Slow food in Tuscany. *Sociologica Ruralis* 42 (4), 312–328.

Mohr, L.A., Webb, D.J. and Harris, K.E. (2001) Do consumers expect companies to be socially responsible? The impact of corporate social responsibility on buying behaviour. *Journal of Consumer Affairs* 35 (1), 45–72.

Nevison, J. (2008) Impacts of sustainable consumption choices on quality of life: The Slow Food example. Unpublished MA (International Leadership) project, Simon Fraser University.

Norwegian Ministry of the Environment (1994) *Oslo Roundtable on Sustainable Production and Consumption*. Oslo: Norwegian Ministry of the Environment.

Nosi, C. and Zanni, L. (2004) Moving from 'typical products' to 'food-related services': The Slow Food case as a new business paradigm. *British Food Journal* 106 (10–11), 779–792.

OECD (2002) *Towards Sustainable Household Consumption? Trends and Policies in OECD Countries*. Paris: OECD.

Osborne, L. (2001) The year in ideas: A to Z: Slow food. *The New York Times Magazine*, 9 December, p. 100.

Parkins, W. and Craig, G. (2006) *Slow Living*. London: Berg.

Peattie, K. and Crane, A. (2005) Green marketing: Legend, myth, farce or prophesy? *Qualitative Marketing Research: An International Journal* 8 (4), 357–370.

Petrini, C. (2001) *Slow Food: Collected Thoughts on Taste, Tradition, and the Honest Pleasures of Food*. White River Junction, VT: Chelsea Green.

Petrini, C. (2007) *Slow Food Nation. Why Our Food Should be Good, Clean, and Fair*. New York: Rizzoli International.

Pietrykowski, B. (2004) You are what you eat: The social economy of the Slow Food movement. *Review of Social Economy* 62, 307–321.

Roberts, J.A. (1993) Sex differences in socially responsible consumers' behaviour. *Psychological Reports* 73, 139–148.

Schlosser, E. (2002) Foreword. In C. Kummer, *The Pleasures of Slow Food: Celebrating Authentic Traditions, Flavours and Recipes* (pp. 10–11). New York: Chronicle Books.

Singh, G. (2005) Slow down and eat. Metroactive – Online document: http://www.metroactive.com/papers/metro/05.01.03/slowfood-0318.html

Slow Food (2010) Online document: http://www.slowfood.com/

Slow Food Foundation (2010) Online document: http://www.slowfoodfoundation.org.

Slow Food in the Canteen (2010) Slow Food in the Canteen: A European school network – Online document: http://dreamcanteen.ning.com/

Terra Madre (2010) Online document: http://www.terramadre.org/pagine/

Van der Meulen, H.S. (2008) The emergence of Slow Food. In W. Hulsink and H. Dons (eds) *Pathways to High-tech Valleys and Research Triangles: Innovative Entrepreneurship, Knowledge Transfer and Cluster Formation in Europe and the United States* (pp. 245–247). Dordrecht: Springer.

Webster Jr., F.E. (1975) Determining the characteristics of the socially conscious consumer. *Journal of Consumer Research* 2, 188–196.

Weigley, E.S. (1974) It might have been euthenics: The Lake Placid conferences and the home economics movement. *American Quarterly* 26 (1), 79–96.

White, B. (2008) The challenges of eating 'Slow'. *The Wall Street Journal*, 3 September, http://online.wsj.com/article/SB122022613854086965.html

Zapata, M.J., Hall, C.M., Lindo, P. and Vanderschaeghen, M. (2011) Can community-based tourism contribute to development and poverty alleviation? *Current Issues in Tourism* 14 (8), 725–749. DOI: 10.1080/13683500.2011.559200.

6 Eat Your Way through Culture: Gastronomic Tourism as Performance and Bodily Experience

Fabio Parasecoli and Paulo de Abreu e Lima

The Brazilian coast is frequently chosen as a destination – both by national and international vacationers – for its natural beauty, beaches and cultural attractions. According to the Brazilian Tourism Board, more than six million foreigners visited the country in 2008, with Rio de Janeiro, the North East and the Amazon the most popular destinations (Brasil Turismo, 2011). Like in other parts of the world, the tourism business often does not fully take into account issues of sustainability, environmental impact, or local social dynamics, which often results in visitors generating inequalities and imbalances (Gössling, 2002; Hughey *et al.*, 2004; Lim & Cooper, 2009; Northcote & Macbeth, 2006). However, halfway between Rio de Janeiro and Sao Paulo, in the town of Paraty, a group of local food producers, restaurateurs and media professionals have launched a programme that they define as 'sustainable gastronomy'.

The town presents great potential for this kind of initiative, since it attracts large numbers of tourists, not only for its beaches, but also for its architectural style and its urban design that dates back to the 18th century. According to the local tourism office, around 200,000 tourists visit the town every year, especially in the summer months between November and March (Paraty Convention and Visitor Bureau, 2011). Developed independently from the international Slow Food movement, which places 'eco-gastronomy' among its core values, the Brazilian initiative aims at highlighting the town's culinary traditions and dishes. These food traditions are rooted both in the local *Caiçara* culture, originally based on subsistence fishing, and in the hinterland of the Mata Atlantica, the rainforest rich in unique edible palms, fruits and berries. The budding initiative, without a central organisation or a structured leadership, has developed due to a consensus-building process that advances through ongoing discussions and pilot actions whose goal is to establish economic and cultural opportunities for the local community. Although the sustainability of the project as a whole does not depend mainly on external elements, such as visitors, but rather on its embeddedness in the local community, it also constitutes a unique offer within the landscape of Brazilian tourism.

Using the Paraty example as a case study, this chapter will examine the concept of 'sustainable gastronomy' (from now on, SG) and its possibilities for the travel industry. In particular, the chapter will explore the potential of SG to ensure the sustainable growth of local communities and to encourage a 'slow' approach to tourism, which would allow visitors to enjoy and participate in food production and culinary traditions as embedded and embodied performances of living cultures. The information has been gathered through bibliographical research, a visit to the area by both authors, personal interviews and emails, and participation in the initiative 'list-serv'.

Food Identity as Tourist Attraction

A new interest in culinary traditions, local products and artisanal delicacies – both national and international – is reaching new heights in Western Europe, Japan, and more recently in the United States and Australia (Amilien, 2005). Other countries are following right behind, such as Brazil, Mexico and Costa Rica, where limited but growing upper classes with disposable incomes have recently shown a shifting sensitivity about the cultural relevance of food traditions (Cafferata & Pomareda, 2009; Granados & Alvarez, 2002). Until a few years ago, many citizens in developing countries would have considered *terroir* ingredients and dishes embarrassing and uncouth, as uncomfortably close to the rural realities and ethnic groups that had often been at the margin of national projects. More recently, growing numbers of consumers are learning to appreciate the role of local communities and their traditions, the manual skills and the know-how of food producers and their ties with historically determined material cultures and the places where they flourished. This trend – quite visible in supermarkets, restaurants and around dinner tables – is promoted and exploited by the media, marketers and politicians and it is also inevitably reflected in tourism.

However, culinary heritage, as an expression of a specific community, is composed not only of ingredients and dishes, but also of bodily practices and living performances passed on through those that the anthropologist Marcel Mauss defined as 'body techniques': gestures, movements, positions that define us both as individuals and as communities (Mauss, 1973). In this sense, traditional food, through the performative and shared aspects of its production, preparation and consumption, is conceptually similar to activities such as dance or song, which can be protected as part of the intangible cultural heritage as defined in the 2003 UNESCO Convention for the Safeguarding of Intangible Cultural Heritage. We can find useful analytical tools to understand these material aspects of community life in embodiment theory, which from the 1980s has underlined the interconnectedness of bodies that extend to interact with other bodies beyond their perceived limits (Mol, 2002). According to this theoretical approach, when we consider bodies in their finite individuality and their completeness, rather than in their relational aspects, we reinforce the concept of the separation between the natural and the social that somehow reproduces the opposition between body and mind, the involuntary and the voluntary. As Blackman (2008: 5) argues, 'we need to move beyond thinking of bodies as substances, as special kinds of things or entities, to explore bodies as sites of potentiality, process, and practice'. These are feeling, affective and eating bodies that are permeable to the outside. After all, food preparation and consumption constitute the perfect example of this inherent permeability and can provide tourists with different experiences. Rosi Braidotti (2002: 70) uses the metaphor of 'nomadic subjectivity', referring to not only ever-changing subject positions but also to bodies considered as not-one, mixing and interconnecting, always morphing, in process, transient and mobile.

Precisely because of their uniqueness and their value in a market that is always looking for the 'next hot thing' and for fresh experiences, traditional foodways can easily be turned into marketable goods and attractions for tourists. Gastronomy and food in the context of travel have been the object of growing interest in academia (Hall *et al.*, 2003; Hjalager & Richards, 2002; Long, 2004; Smith & Costello, 2009). The tourism industry, always in a quest for new and diverse destinations, has turned culinary traditions, especially those embedded in local customs and culture, into conspicuous elements of attraction for visitors, reshaping the industry's perceptions about the elements that constitute a trip or a vacation.

As a consequence, recent discourses and practices around food and tourism have been frequently constructed around interpretive frames that refer to values such as authenticity, exoticism and cosmopolitanism, in part appealing to the tourist's desire to acquire cultural capital (Germann Molz, 2007; Johnston & Baumann, 2007). The success of food-focused destinations often depends on factors such as locality, history, identity, traditional foods and foodways – and their growing relevance in tourism. However, these places cannot remain isolated from the dynamics of globalisation that inform many aspects of contemporary material culture. Local food-related practices and productive systems are often fragile and can be threatened by excessive exposure to foreign travellers and the needs of the tourism industry to minimise expenses, maximise gains and streamline operations (Belisle, 1984). By taking into consideration these risks and constructing new forms of interaction between visitors and their destinations, 'slow' travel practices could establish modalities that allow local foodways to become viable sources of revenue for the local communities without losing their social and cultural relevance.

Many communities, especially in the coastal areas of less developed countries, have suffered because of solidification of so-called 'tourist bubbles' that limit the actual interactions between tourists and the local environment, preventing visitors from having access to food establishments and products outside the area deemed safe for their own good (Carrier & Macleod, 2005; Jaakson, 2004; Jacobsen, 2003). Participation of developing communities in worldwide flows of people, money, goods and information is not necessarily and univocally a cause of progress, but depending on local social and political factors may cause at times disruptions and tensions. As a consequence, the political and ideological discourses and practices around the defence of traditions, customs, dishes and products are often created within a dichotomy that opposes not only global versus local, but also homogeneity versus diversity and universality versus particularity. The same oppositions emerge in a marketing approach often adopted by the tourism sector, which presents local foodways as unadulterated remnants from the past that hold their ground against modernisation and widespread loss of cultural identity. Both for tourist developers and the defenders of traditional food, although the motivations might differ, the immediate goal is to underline the uniqueness and the cultural value of specific food experiences and to defend them against the global tide of standardisation.

This set of oppositions, however, is clearly an oversimplification. Since both the local and the global are socially and historically constructed, it is necessary to abandon the naïve point of view that considers the local as 'natural', original, connected to biodiversity and heterogeneity, and as the last defence against the homogenising, unnatural forces of globalisation (Hardt & Negri, 2001). Many social movements have shown that it is often necessary to fight globally to defend the local. Furthermore, when it comes to food, it is arguable that in many cases local identities are inherently dialectic, as they are the result of larger trade and contact networks, and are acknowledged and defined as local precisely against the backdrop of other, distant places. Places like Brazil and the Caribbean are good examples of these historical dynamics. Central and South American former colonies attracted waves of immigrants and economic actors that each left a visible mark in the local foodways. At times, as Richard Wilk (2006) aptly illustrates in his analysis of the formation of Belizean national cuisine, it is only when nationals from these areas move to countries with identifiable and culturally acknowledged food traditions that they realise the need to define and codify their own culinary identity.

From a practical point of view, international exposure can lead to increased demand and prices, which can revamp or even save failing ingredients or dishes, as prominent initiatives like the Slow Food *praesidia*, amplified through social actions, media campaigns and political interventions, have demonstrated (Leitch, 2003; Parasecoli, 2003). However, as Barbara Kirshenblatt-Gimblett (1995: 369) has argued, the identification of heritage is 'a mode of cultural production in the present that has recourse in the past ... (it) is a value added industry ... (it) produces the local for export'. Revitalising (or even resurrecting) a culinary tradition is not only an operation on the past but also can be a measure to solidify the present and guarantee a better future for the communities involved, who are in part defined by their present-day dynamics including their relationship with global tourism. Carlo Petrini (2001: 97), the founder of the Slow Food movement, has pointed out the importance of grafting local projects onto global consumption networks by dynamics that pair 'ancient and marginal professions and a new class of consumers disposed to pay a fair price in exchange for quality and outstanding flavor', while keeping in mind issues of sustainability and social justice. The products, skills and daily practices at the centre of culinary revival and promotion projects often depend for their financial viability on the existence of consumers and tourists elsewhere. However, it would constitute a safer and more sustainable objective for the communities involved to anchor the success of their initiatives first and foremost in the community itself, as the SG initiative in Paraty aims at doing.

Paraty and the Sustainable Gastronomy Initiative

Within the cultural, social and economic framework that turns local and traditional foodways into an attraction, the very notion of authenticity, often invoked as a motivation and as a marketing tool both in the food industry and in the tourism industry, becomes fraught with all kinds of contradictions. Since the valorisation of local products and practices is actually a social process embedded in specific contemporary realities, authenticity cannot be limited to either a philological repetition of the past, or protected from contaminations determined by the forces of globalisation. Food skills, preparations and productions, when presented to tourists in museum-like structures or in well-designed and precise historical re-enactments, risk losing much of their appeal. In this way, visitors may be prevented from becoming not only spectators, but also participants, ultimately denying their embodied connection with eating and consumption. A well-preserved and picturesque urban structure like Paraty could easily lead local actors to use it as a set to recreate situations and environments inspired by the colonial past of the town.

In this location, however, many food businesses have chosen a different path to dialogue with the local culinary traditions that are often referred to as 'Caiçara cuisine'. The Caiçara communities originated in the 16th century through the miscegenation of white Europeans, mostly of Portuguese origin, with the native tribes settled along the seacoast from the south of the state of Rio de Janeiro to the north of the state of Paraná. Later on, black Africans from urban areas also migrated to the coasts and contributed to the array of hybridised local cultures. The area is rich with fresh ingredients such as native scallop varieties, fish and produce from the Atlantic rainforest. *Caiçara* culture reflects the realities of the local mixed-race communities, who slowly slipped into decadence after a period of relative splendour connected to the commercial activity of the port, through which the gold from the interior transited before Rio de Janeiro acquired political and economic pre-eminence (Gurgel & Amaral, 1973). As a consequence, the customs there expressed the need for a very practical and substantial type of food, which from the start could not embrace the basis of European Mediterranean cuisine such as wheat and grapes, among others, for these ingredients could not find a proper climate for cultivation in the region. Slowly, other ingredients were incorporated: *pirão* (a thickened fish broth), toasted ground manioc, roots, fish and game meat, banana and corn and more. As a consequence, the *Caiçara* cannot be understood as an exclusively fishing culture. There can be enormous differences in daily habits between communities that live on the beach and those located on the hillside (Bueno *et al.*, 2007; Peçanha, 2010).

Following the budding but fast growing interest in food traditions, *Caiçara* cuisine is also experiencing a revival, with the town of Paraty and the SG initiative as a laboratory for culinary experiments that fuse 21st century service and techniques to traditional produce and dishes. Restaurants such as the refined Banana da Terra, Casa do Fogo – where food is often flambé with the local specialty *cachaça* – Alquimia dos Sabores and La Luna, just to mention a few, have spearheaded the rediscovery of local food. Furthermore, the SG participants underline the crucial role of alternative circuits besides restaurants to avoid the formation of closed networks and to be permeable to other elements in the local community, such as producers, institutions such as schools, and private consumers.

Inevitably, an element of romanticism has emerged in the discourse surrounding the *Caiçara* cuisine, often making the harsh realities of the contemporary *Caiçara* communities less visible. As Brazilian author Vito d'Alessio states in his introduction to the first illustrated, full colour coffee table book about *Caiçara* cuisine, written in both Portuguese and English:

> It is a different kind of sophistication which does not pass through alchemy but rather with the possibility of dealing with fish, game, and planting to extract from nature and all its cycles what is essential to maintain a simple and cozy lifestyle, in one of the most beautiful and harmonious regions of the planet. It is in this simplicity, that a great distinction of caiçara cuisine resides, a genuine way of understanding cooking. (cited in Bueno *et al.*, 2007: 7)

The references to the somehow innate culinary knowledge of the locals, always attuned to their natural environment and enjoying a humble but comfortable lifestyle, are a good example of this kind of romanticism.

Local producers, restaurateurs and media professionals in Paraty have started working together in a project that aims to 'research, develop, and diffuse a cuisine with international standards, inspired by the local culture and the utilization of organic agroforestry, livestock, and fishery products, ecologically produced by local communities' without denying the commercial and economic downfall of the new interest towards the local cuisine (Gastronomia Sustentável de Paraty, 2010). The members of the association that is promoting the SG project agreed on the following objectives and policies that:

(1) Prefer local products from family-based agriculture that do not degrade farmers, consumers, or the environment.
(2) Prefer the products of the sea, artisanal fish or those produced in sea farms by local fishermen, respecting the times of moratorium.
(3) Indicate the provenance of products on the menus.
(4) Stimulate the proximity and the interaction among tourists, producers, and communities.
(5) Separate used vegetable oils from the garbage to be recycled, and avoid the use of non-recyclable packaging.
(6) Use grease traps and install, over time, biological filtering systems to treat used water and sewage.
(7) Manage the use of energy to reduce consumption.
(8) Sustain and articulate the process of local certification for producers.
(9) Work together with local farmers with the goal to have public powers guarantee parcels of land for family agriculture, through the legalization of land ownership.
(10) Monitor and assess the process through reliable indicators. (Gastronomia Sustentável de Paraty, 2010)

It may sound like a very ambitious set of goals, but the participants remain pragmatic. In regular meetings they have been outlining various practical actions that they can initiate without changing their ongoing activities or the business models already in place. In August 2010, on the occasion of the Festa Literária Internacional de Paraty (FLIP), a world-renowned literary festival that attracts significant crowds every year, a parallel series of events, known as Off-Flip, were organised. According to the Paraty tourism association, in 2008 more than 20,000 visitors participated in the festival (Paraty Tur, 2011). Its purpose was to highlight local values and not only the spectacular and commercial elements that might attract many visitors. Embracing the Off-Flip philosophy, some of the participants of the SG initiative arranged a group called Circuit of Literary Dishes of the Sustainable Gastronomy of

Paraty in which each restaurant created a limited-edition dish or mixed drink inspired by famous Brazilian authors, from the sociologist Gilberto Freire to the novelist Jorge Amado.

Although the initiative is not directly under the control of the local political and administrative authorities and is rather the result of ongoing interactions within a network of local actors, it is supported by Agência de Apoio ao Empreendedor e Pequeno Empresário (SEBRAE, the Agency of Support to Entrepreneurs and Small Businesses) of the state of Rio de Janeiro, under the general, hands-off observation of the Ministry of Agriculture (MAPA, Ministerio da Agricultura, Pecuaria e Abastecimento). The relevance of tourism for this initiative is evident. Although local producers and entrepreneurs aim at creating synergies based on a solid network of local actors in order to establish a form of SG based on local traditions, they also take into consideration the presence of tourists, who are considered not only as possible buyers but as involved actors. Clearly, the SG approach can create an optimal interaction with tourists who share the same basic priorities, can appreciate the value of unique ingredients, dishes and practices, and are ready to invest disposable income and cultural capital to enjoy them. The visitors that fill the streets of Paraty for the FLIP and Off-Flip events are actually the perfect target segment of consumers.

The Case of the Local *Cachaça*

The literary festival is not the only local event that attracts visitors from Rio de Janeiro and São Paulo, and that puts Paraty on the map as a tourist destination. In the month of August, the town organises the Festival da Cachaça (the Cachaça Festival) which highlights one of the most traditional products of the area (Paraty Turismo e Ecologia, 2009). Since the period when Brazil was a Portuguese colony, the town has been known as a centre for the production of spirits made from sugar cane, known as 'cachaça'. *Cachaça* production is deeply rooted in historical, cultural, social and economic aspects of the region. The caravans that brought gold from the state of Minas Gerais in the interior along the so-called Caminho do Ouro (the Path of Gold), the route connecting the mines to the coast, travelled back carrying barrels of the liquor. The tradition never died completely, even when the caravans started heading towards Rio de Janeiro and the city fell into a period of economic decadence. The product has recently been revitalised by local entrepreneurs, some of whom still use the old water mills to squeeze the sugar cane. Finally, in 2007, the Association of Local Artisan Cachaça Producers (APACAP) obtained from the National Institute of Industrial Property (INPI), in cooperation with the Ministry of Agriculture, its recognition of Geographical Indication (GI) as 'Cachaça de Paraty'.

According to Raul Bittencourt Pedreira from the INPI, *cachaça* producers had been trying to obtain an official seal from the Brazilian government since 2002 in order to protect good quality *cachaça* from cheap counterfeit products (personal communication, interview with Raul Bittencourt Pedreira, 20 August, 2011). Apparently the effort had not been successful because it was difficult to distinguish among production regions, local cultures and water and raw materials that were used. In the beginning of 2007, however, following the guidelines for improvements in local production and distillery methods issued by the Ministry of Agriculture, the project of Geographical Indication for the Paraty *cachaça* was properly presented to the INPI. At the time there were only a few producers, most of them being medium rural producers and entrepreneurs. The entire bureaucratic process for the GI approval took less than a year, which has been perceived as extremely fast by the parties involved and interpreted as a sign of interest both from the local producers and the national authorities to promote the spirit and the area. In the past, many people in Brazil believed in the tale that good *cachaça* could not be produced near the ocean because the marine breeze interfered with the production process. Contrary to this belief, this area's award-winning *cachaças* are produced very close to the water.

One of the difficulties that the industry is experiencing today is the supply of raw material, due to the growing success of the product. The local sugar cane growers are not yet included in the GI regulation, which only focuses on the production and characteristics of the spirit. The current directive determines that the *cachaça* producers should slowly stop using sugar cane that grows outside of the GI area and only buy locally grown raw materials. This would put the local growers in a stronger negotiating position and the benefits reaped from the GI would be redistributed more equitably along the production chain. To this day, however, there are neither deadlines nor penalties for continuing the usage of ingredients from other areas.

The GI cachaça producers are showing a clear interest in tourism as a possible source of revenue. Some of the producers are already making changes to their facilities in order to welcome visitors, allowing them to see the various phases of production and to buy spirits on the premises. They also actively participate in the projects related to the maintenance and development of the historical sites and natural environment along the Caminho do Ouro, hoping to diversify the local tourism offered (Centro de Informações Turísticas Caminho do Ouro, 2010). *Cachaça* producers are particularly involved in the Festival da Cachaça, which until a few years ago was actually called Festival da Pinga. *Pinga*, a popular name for the spirit, is now perceived to be too vulgar and connected to a past of poverty and alcohol abuse, potentially tainting the image of a high quality product deserving a GI. The Festival da Cachaça showcased several stands of products and producers involved in the SG initiative, showing a coordinated effort among all local actors to promote local traditions. So far, the *cachaça* producers have not managed to convey the relevance of the GI recognition to the community, to the point that most citizens of Paraty are not even aware of its existence. This lack of communication and the problems that it creates have prompted the SG participants to emphasise the relevance of community involvement in the initiative, without relying excessively on tourists.

Developing a New Model of Slow Tourism

The SG initiative gives priority to local networks in order to ensure viable and lively businesses. However, it cannot be denied that the various tourism events and the crowds they attract to Paraty help the SG participants to thrive. As the fourth point of the SG principle states, the project aims at overcoming any excessive separation between service providers and visitors by involving them more directly in the finalities of the overall project. Food provides a suitable medium for this kind of interaction, not only because eating and ingestion engages the dimension of physicality as a fundamental component of the travel experience, but also because visitors are called on to assume the role of co-producers. The concept of 'eating as an agricultural act', introduced by Wendell Berry and embraced by Slow Food, could remind tourists of the embodied aspects of food production and their responsibilities in terms of environment and social justice (Berry, 1992; Cameron & Gibson-Graham, 2003; Staeheli et al., 2004). This approach might offer a wider and more holistic form of participative tourism than, for instance, ecotourism, at least in those instances where entrepreneurs external to the community organise the activities for the visitors and locals are exclusively involved as service providers, without any engagement in the decision-making process regarding the use and potential of the land.

Furthermore, in emphasising the crucial role of the local community, SG does not aim at turning local products and traditions into museum objects that cannot dialogue with the current environment and contemporary dynamics lest they lose their veneer of authenticity. SG considers ingredients, dishes and practices as part of a living tradition that cannot be isolated by larger issues of community development, including tourism. The main goals are to ensure a decent livelihood for the dwellers of the town and its vicinities in the long term, respect the environment and allow all actors to have a say in common decisions. If authenticity were instead connoted in terms of purity and abstraction from the messiness of the present, it could easily end up downplaying the intense and embodied social investments in place and the locality that underlie the identification, revival and promotion of food products and traditions. This would lead to preventing the formation of action-based groups that may function as catalysts for the development of a new sense of community, like in the case of the participants in the SG initiatives in Paraty. These groups and their connection to a specific environment – which requires that attention be paid to material elements such as food, ingestion and the body – can help counter-balance the detachment from corporeality that paradoxically is detectable in many tourism initiatives, which in principle should enhance the travellers' experience in all its aspects. Too often, safety concerns on one side and the desire to meet the media-influenced expectations for pre-packaged authenticity and uniqueness – as moulded by the global media – on the other can leave tourists parading in front of lifeless and sanitised adaptations of otherwise lively and embodied realities.

These preoccupations can lead to the dynamics often visible in resorts, where local dishes are either not offered at all or are tamed to satisfy the supposed bland palate of the average tourist, while right outside the gates stalls and little restaurants might be peddling the traditional fare, usually less domesticated than the resort version but still appealing to a wide range of visitors, at much lower prices. Paradoxically, the search for authenticity might prevent tourists and consumers in general from gaining access to reality, as studies about *terroir* and the connection between place, heritage and food have indicated (Altaffer, 2009; Ilbery *et al.*, 2005; Tellström *et al.*, 2006). While ecotourism invites visitors to develop a sensibility for the local environment (Dinan & Sargeant, 2000; Reddy, 2008), culinary tourism and initiatives like SG have the potential to allow them to blur the boundaries that separate them from that very environment. After all, when we eat we ingest the outside world and we allow it to enter us. If we eat local products, we make the places we are visiting a part of ourselves. It goes without saying that any bubble approach to tourism denies tourists the possibility of this experience.

Experiments like SG in Paraty instead suggest the viability of forms of 'slow' tourism. But how can we interpret this 'slow' quality? Parkins and Craig (2006: 5) have defined 'slow living' as,

> a process whereby everyday life – in all its pace and complexity, *frisson* and routine – is approached with care and attention, as subjects attempt to negotiate the different temporalities that they daily experience. It is above all an attempt to live in the present in a meaningful, sustainable, thoughtful, *and pleasurable* way ... what this requires is learning a 'new consciousness practice', involving the body and emotions as well as perception and thought.

However, as Parkins and Craig point out, slow living cannot be countercultural in that it does not aim at establishing new and alternative spaces, but rather operates with the goal of changing everyday realities. This approach is the basis for what Slow Food (2010) defines as 'eco-gastronomy':

> We believe that everyone has a fundamental right to pleasure and consequently the responsibility to protect the heritage of food, tradition and

culture that make this pleasure possible. Our movement is founded upon this concept of eco-gastronomy – a recognition of the strong connections between plate and planet. Slow Food is good, clean and fair food. We believe that the food we eat should taste good; that it should be produced in a clean way that does not harm the environment, animal welfare or our health; and that food producers should receive fair compensation for their work. We consider ourselves co-producers, not consumers, because by being informed about how our food is produced and actively support-ing those who produce it, we become a part of and a partner in the pro-duction process.

Emphasis is thus given to sustainable food systems, biodiversity, ethical issues and environmental and social contexts. The SG initiative in Paraty, despite not being primarily geared towards tourism, seems to embrace many of these concepts. By trying to engage visitors in sustainable long-term com-munity development, it inherently invites them to embrace a different approach to their presence in the local environment. And by deciding to patronise a business promoting SG, tourists can be made aware that they are not only taking advantage of a service, but also participating in the revival and the growth of culinary traditions and practices that other-wise might be in danger of extinction. This opportunity might appeal to those who practice various forms of food-related ethical consumption, from fair trade to community sustained agriculture, in order to participate in the social and political debates that surround them (Clarke *et al.*, 2007; Goodman, 2004).

Conclusion

For the tourism industry, food has become a crucial element in determin-ing and shifting customer flows. Local traditions and products often constitute valuable assets in defining the attractiveness of a destination. However, we cannot forget that the practical and artisanal dimension of gastronomy origi-nates in the practices and the know-how of food producers and restaurateurs, rooted in the lived environment and the customs of their communities.

To avoid the risk of turning these cultural elements into a spectacle, per-formed by locals merely for the benefit of tourists, the development of action-groups aiming to promote local gastronomy needs to be based on dialogues and interactions solidly embedded in the daily aspects of community life. Food-related activities and the network of social and cultural practices that revolve around them remind us that all actors involved are actually embodied and participate in them not only in spirit but also in the flesh (Valentine, 2002).

The SG project in Paraty, as implemented by the local stakeholders and embedded in the daily existence of their community, highlights the protection of embodied and living practices as valuable heritage. At the same time, the undertaking is solidly rooted in the realities of the shifting and globalised contemporary culture. It does not codify, ossify and turn local culinary traditions into museum pieces that cannot evolve and that, as a consequence, could wither. Since the SG initiative does not limit itself to intellectual or consumerist approaches that inherently point to exploitation and distance between customers and producers, it has the intrinsic potential to entice tourists to discover and intermingle creatively with the material life and the food-related experience of the area they visit. The Paraty SG experience, albeit still in its early phases, can contribute to developing a different model of slow travel and slow tourism that highlights food, not just as another attraction to be quickly (and mindlessly) consumed, but rather as the core of productive and satisfying interactions between local communities and temporary visitors.

Acknowledgements

The research for this essay was funded by a grant from the University of Gastronomic Sciences, Italy. Special appreciation goes to Gilberto Mascarenhas (MAPA/RJ and Research Group 'Market, Network and Values') and all the participants in the Sustainable Gastronomy project in Paraty.

References

Altaffer, P. (2009) Regional branding: Exploring a ripe marketing opportunity for the natural products industry from the corners of the world. *Nutraceuticals World* 12 (6), 6.
Amilien, V. (2005) Preface: About local food. *Anthropology of Food*, 4 May – Online document: http://aof.revues.org/document305.html
Belisle, F.J. (1984) Tourism and food imports: The case of Jamaica. *Economic Development and Cultural Change* 32 (4), 819–842.
Berry, W. (1992) The pleasures of eating. In D.W. Curtin and L.M. Heldke (eds) *Cooking Eating Drinking: Transformative Philosophies of Food* (pp. 374–381). Bloomington; Indianapolis: Indiana University Press.
Blackman, L. (2008) *The Body*. Oxford: Berg.
Braidotti, R. (2002) *Metamorphoses: Towards a Materialist Theory of Becoming*. Oxford: Polity Press.
Brasil Turismo (2011) Turismo no Brasil – Dados do Turismo [Tourism in Brazil – Tourism Data] – Online document: http://www.brasilturismo.com/turismo/dadosdoturismo.php
Bueno, A., Diegues, A.C. and D'Alessio, V. (2007) *Culinária Caiçara: O Sabor entre a Serra e o Mar* [*Caiçara Cooking: Flavors between Mountains and Sea*]. São Paulo: Dialeto.
Cafferata, J.P. and Pomareda, C. (2009) *Indicaciones Geográficas y Denominaciones de Origen en Centroamérica: Situación y Perspectivas* [*Geographical Indication and Origin Denomination in Central America: Situation and Perspectives*]. Geneva: ICTSD.
Cameron, J. and Gibson-Graham, J.K. (2003) Feminizing the economy: Metaphors, strategies, politics. *Gender, Place and Culture* 10 (2), 145–157.
Carrier, J.C. and Macleod, D.V.L. (2005) Bursting the bubble: The socio-cultural context of ecotourism. *Journal of the Royal Anthropological Institute* 11 (2), 315–334.

Centro de Informações Turísticas Caminho do Ouro (2010) O Projeto [The Project]–
Online document: http://www.paraty.com.br/caminhodoouro/projeto.htm

Clarke, N., Barnett, C., Cloke, P. and Malpass, A. (2007) Globalising the consumer: Doing
politics in an ethical register. *Political Geography* 26 (3), 231–249.

Dinan, C. and Sargeant, A. (2000) Social marketing and sustainable tourism: Is there a
match? *International Journal of Tourism Research* 2 (1), 1–14.

Gastronomia Sustentável de Paraty (2010) Vamos sanear os nossos quintais [Let's clean
up our backyards]. *Folha Do Litoral* 12 (89), 3.

Germann Molz, J. (2007) The cosmopolitan mobilities of culinary tourism. *Space and
Culture* 10 (1), 77–93.

Goodman, M.K. (2004) Reading fair trade: Political ecological imaginary and the moral
economy of fair trade foods. *Political Geography* 23 (7), 891–915.

Gössling, S. (2002) Human-environmental relations with tourism. *Annals of Tourism
Research* 29 (2), 539–556.

Granados, L. and Alvarez, C. (2002) Viabilidad de establecer el sistema de denominaciones
de origen de los productos agroalimentarios en Costa Rica [Feasibility of establishing
a system of appellations of origin of food products in Costa Rica]. *Agronomía
Costarricense* 26 (1), 63–72.

Gurgel, H. and Amaral, E. (1973) *Paraty: Caminho do Ouro*. Rio de Janeiro: Livraria São José.

Hall, C.M., Sharples, L., Mitchell, R., Macionis, N. and Cambourne, B. (2003) *Food
Tourism Around the World: Development, Management and Markets*. Burlington, MA:
Butterworth-Heinemann.

Hardt, M. and Negri, A. (2001) *Empire*. Cambridge: Harvard University Press.

Hjalager, A. and Richards, G. (eds) (2002) *Tourism and Gastronomy*. New York:
Routledge.

Hughey, K., Ward, J., Crawford, K., McConnell, L., Phillips, J. and Washbourne, R. (2004)
A classification framework and management approach for the sustainable use of
natural assets used for tourism. *International Journal of Tourism Research* 6 (5),
349–363.

Ilbery, B., Buller, H., Kneafsey, M., Morris, C. and Maye, D. (2005) Product, process and
place. An examination of food marketing and labeling schemes in Europe and North
America. *European Urban and Regional Studies* 12 (2), 116–132.

Jaakson, R. (2004) Beyond the tourist bubble? Cruiseship passengers in port. *Annals of
Tourism Research* 31 (1), 44–60.

Jacobsen, J. (2003) The tourist bubble and the Europeanisation of holiday travel. *Journal
of Tourism and Cultural Change* 1 (1), 71–87.

Johnston, J. and Baumann, S. (2007) Democracy versus distinction: A study of omnivo-
rousness in gourmet food writing. *American Journal of Sociology* 113 (1), 165–204.

Kirshenblatt-Gimblett, B. (1995) Theorizing heritage. *Ethnomusicology* 39 (3), 367–380.

Leitch, A. (2003) Slow food and the politics of pork fat: Italian food and European
identity. *Ethnos* 68 (4), 437–462.

Lim, C. and Cooper, C. (2009) Beyond sustainability: Optimising island tourism development. *International Journal of Tourist Research* 11 (1), 89–103.

Long, L. (ed.) (2004) *Culinary Tourism*. Lexington: The University Press of Kentucky.

Mauss, M. (1973) Techniques of the body. *Economy and Society* 2 (1), 70–85 (original work
published 1935).

Mol, A.M. (2002) *The Body Multiple: Ontology in Medical Practice*. London; New York: Duke
University Press.

Northcote, J. and Macbeth, J. (2006) Conceptualizing yield: Sustainable tourism management. *Annals of Tourism Research* 33 (1), 199–220.

Parasecoli, F. (2003) Postrevolutionary chowhounds: Food, globalization and the Italian
left. *Gastronomica* 3 (3), 29–39.

Paraty Convention and Visitor Bureau (2011) Paraty C&VB – Online document: http://www.paratycvb.com.br/site/paraty-cvb/

Paraty Tur (2011) *Flip 2009 movimenta cidade histórica no Rio* [*Flip 2009 Enlivens Historical City in the State of Rio*] – Online document: http://www.paraty.tur.br/noticias/ler.php?id = 4

Paraty Turismo e Ecologia (2009) XXVIII Festival da Cachaça, Cultura e Sabores de Paraty [Festival of Cachaça, Culture, and Flavors of Paraty] – Online document: http://www.paraty.com.br/feriados/festivaldapinga/

Parkins, W. and Craig, G. (2006) *Slow Living*. Oxford: Berg.

Peçanha, G. (2010) *Delícias de Paraty [Delights of Paraty]*. Sao Paulo: Baraúna.

Petrini, C. (2001) *Slow Food: The Case for Taste*. New York: Columbia University Press.

Reddy, M.V. (2008) Sustainable tourism rapid indicators for less-developed islands: An economic perspective. *International Journal of Tourism Research* 10 (6), 557–576.

Slow Food (2010) Our philosophy – Online document: http://www.slowfood.com/about_us/eng/philosophy.lasso

Smith, S. and Costello, C. (2009) Segmenting visitors to a culinary event: Motivations, travel behavior, and expenditures. *Journal of Hospitality Marketing and Management* 18 (1), 44–67.

Staeheli, L., Kofman, E., and Peake, L. (eds) (2004) *Mapping Women, Making Politics: Feminist Perspectives on Political Geography*. London: Routledge.

Tellström, R., Gustafsson, I. and Mossberg, L. (2006) Consuming heritage: The use of local food culture in branding. *Place Branding* 2 (2), 130–143.

Valentine, G. (2002) In-corporations: Food, bodies and organizations. *Body and Society* 8 (2), 1–20.

Wilk, R. (2006) *Home Cooking in the Global Village*. Oxford; New York: Berg.

7 'Make Haste Slowly': Environmental Sustainability and Willing Workers on Organic Farms

Margo B. Lipman and Laurie Murphy

In spite of occasional shocks, international tourist arrivals have shown virtually uninterrupted growth – from 25 million in 1950, to 277 million in 1980, to 438 million in 1990, to 681 million in 2000, and the current 880 million. (World Tourism Organization (UNWTO), 2010: 1)

The tourism industry is quickly recovering from the Global Financial Crisis. For example, from January to June 2010 worldwide international tourist arrivals grew by an impressive 7% (UNWTO, 2010). Growth is expected to continue, as from 2009 to 2020 international arrival figures are projected to nearly double to 1.6 billion travellers (UNWTO, 2010). Domestic tourism volumes are also expected to increase, with particularly rapid growth in countries such as China and India (Scott *et al.*, 2008). The proportion of trips that are long-haul, as well as overall average trip distance, are also likely to increase significantly in the coming years (Scott *et al.*, 2008). Furthermore, the proportion of travellers arriving at their destination by air has been steadily growing (53% in 2009) (UNWTO, 2010).

In the following chapter we argue that current and future (projected) patterns of tourism are environmentally unsustainable. One way in which this situation can be ameliorated is through a change in consumer behaviour. The suggestions outlined below are largely related to transport emissions and coincide with the tenets of slow tourism. Slow tourism, however, may be challenging in some contexts. Work exchange programmes, such as Willing Workers on Organic Farms (WWOOF), may help travellers to overcome some of the obstacles to slow tourism, which include accessing rural areas using shared surface transport (coach, trains) and engaging with locals beyond those who work in the tourism industry, as well as facilitate a more sustainable approach to tourism among those not explicitly interested in 'slowness'. Unfortunately, there is a dearth of research on both Willing Workers on Organic Farms and work exchange programmes in general. The research presented in the following chapter aims to help address this gap in knowledge and understanding, with a focus on the connections between slow tourism and the WWOOF programme.

Tourism and Climate Change

Tourism used to be called a 'smokeless' industry because it was not considered to be harmful to the environment. Although this misconception may still be held by the uninformed, most people familiar with the industry now acknowledge that this is far from the truth. Tourism causes a great deal of harm to the environment and, unfortunately, the impacts are getting worse, not better (Hall, 2009). The negative environmental effects associated with tourism include pollution, changes in land use and cover, resource use, biodiversity loss and disruption of ecosystems (Gössling, 2002; Hall, 2009). Likely, the most important impact, however, is tourism's contribution to anthropogenic global warming.

In a technical report for the World Tourism Organization and United Nations Environment Programme, Scott et al. (2008) used two different types of measurements to assess the industry's contribution to this phenomenon: carbon dioxide (CO_2) emissions and radiative forcing. Carbon dioxide is the most significant contributor to human-induced climate change; however, other gases play an important role as well (Scott et al., 2008). With regard to aviation, it is also necessary to take into account the additional warming effect emissions cause at flight altitude (Scott et al., 2008). Radiative forcing[1] is thus a more comprehensive measure than CO_2 emissions, as it 'measures the extent to which emissions of greenhouse gases raise global average temperatures' (Scott et al., 2008: 121). It is, however, a more uncertain and complex measurement, so it is worthwhile to also assess levels of CO_2 emissions.

Scott et al. (2008) estimate that tourism's contribution to global CO_2 emissions is about 5%, although the contribution to radiative forcing is significantly higher at 7.8% (as of 2005). Although these figures are significant and worthy of serious mitigation efforts, future trends in tourism are even more worrisome. The technology-adjusted 'business as usual' scenario leads to a 160% increase in CO_2 emissions and about a 200% increase in tourism's contribution to radiative forcing from 2005 to 2035 (Scott et al., 2008). This is due to the expectation that many more people will be travelling, more often, and greater distances (Scott et al., 2008).

Transport to and from the destination is responsible for 75% of tourism's CO_2 emissions and 90% of its contribution to radiative forcing (Scott et al., 2008). More specifically, aviation is currently responsible for the largest single share of both emissions and radiative forcing (40%, 75%), and both of these shares are expected to increase over the coming years to approximately 53% and 84% (Scott et al., 2008). This is largely due to the prediction that airline passenger numbers and distances travelled will grow by 4% and 5% per year, respectively, over the next 20 years (Gössling et al., 2010). Long- and medium-haul flights are the most detrimental, for example, a flight from the United Kingdom to Australia causes around 4.8 tons of CO_2 to be released per passenger (Scott et al., 2008). This is greater than the annual emissions of the average world citizen, and is more than half of the annual emissions of the average European Union (EU) citizen (Scott et al., 2008).

The second largest share of emissions and radiative forcing is from car transport (32%, 14%), while 'other transport' accounts for a negligible share of both (Scott et al., 2008). This is due to fewer people using other forms of transport, but also because rail and coach are substantially more efficient than cars or planes. With regard to CO_2 emissions, the average airplane and car (two passengers) is approximately five times less efficient than rail, and six times less efficient than coach. These figures take into account the average occupancy rate for each mode, which is why coach transport is more efficient than rail (Scott et al., 2008).

Slow Tourism

There are many people who would like to depend on technology to solve all of the world's problems so they can go about their business as usual and avoid any difficult behavioural changes. Nevertheless, we believe it is necessary to act *now* on climate change. We cannot wait for technological solutions as, for example, the most optimistic scenarios for aviation predict a 2% reduction in emissions per year over the next 20 years (Gössling et al., 2010). This is nowhere near a large enough improvement, particularly given the previously discussed trends in tourism. One of the alternative options for reducing tourism's contribution to global warming, and the focus of the following discussion, is a change in consumer behaviour.

The goal of sustainable tourism consumption may be achieved through a 'slower' approach to tourism, with a particular focus on transport. Slow tourism has been conceptualised in different ways, depending on the focus of the author(s). For example, Conway and Timms (2010) stress the potential for slow tourism branding of remote Caribbean areas as an avenue for development. They do not even comment on the environmental implications of relying solely on medium- and long-haul source markets (USA and Europe), particularly given the extra transport required to access the remote locations. In contrast, Dickinson and Lumsdon (2010) focus, as we do in this chapter, on slow tourism and environmental sustainability. It is therefore proposed that *low-carbon* slow tourism has the following key components: reduced (carbon producing forms of) mobility, a shift in the type of transport chosen and engagement with local communities.

Reduced mobility involves people taking fewer trips, which may be of longer length, and eschewing long-haul destinations (Dickinson & Lumsdon, 2010; Hall, 2009). Domestic, regional and local travel – discovering what is in your own back yard – is particularly encouraged (Dickinson & Lumsdon, 2010). Also important is choosing to visit fewer destinations and instead spending more time in each location (Hall, 2006; Matos, 2004; Germann Molz, 2009). Lastly, travelling less within each destination is an integral element, and this includes avoiding carbon-intensive activities (Dickinson & Lumsdon, 2010).

As can be surmised from the previous section, careful consideration of the *type* of transport chosen is essential to low-carbon slow tourism. Self- or animal-propelled is ideal (e.g. horses, bicycle), and avoidance of airplanes, cruise ships and cars[2] is a high priority (Dickinson & Lumsdon, 2010). Coaches, trains, motorbikes and some types of boats are more responsible (and often slower) options (Dickinson & Lumsdon, 2010). Modal choice is important when considering travel to the destination(s), as well as travel within each destination (Dickinson & Lumsdon, 2010). For example, it is preferable to walk or ride a bicycle to visit attractions at a destination rather than use a private car.

Engagement with and exploration of local communities (people, places and culture) is the final component of slow tourism (Dickinson & Lumsdon, 2010). Although some aspects relate to environmental sustainability, the key focus with this element is immersion and participation, instead of simple observation, at an unhurried, leisurely pace (Germann Molz, 2009). Visiting fewer destinations and staying longer in each location can assist slow tourists in their search for quality, deep and rich experiences instead of trying to see *as much as possible* superficially (Conway & Timms, 2010; Knight, 2010). The slow tourist should be attentive to immersing oneself in the local heritage and culture, '... establishing local routines, indulging in local cuisines, and becoming connoisseurs of the local culture' (Germann Molz, 2009: 280). Appreciating any local natural resources is also important, as well as frequenting locally-owned, sustainable businesses.

In addition to reducing the environmental impact of one's trip, taking a 'slow' approach to tourism has several important benefits for both the tourist as well as the communities visited. The traveller will have the opportunity to develop a deeper understanding of, and relationship with, the people, places and cultures visited (Knight, 2010; Matos, 2004; Germann Molz, 2009). This type of tourism will also encourage the retention of local distinctiveness and a sense of *place*, in opposition to the worldwide trend towards homogenisation (Conway & Timms, 2010). This potential benefit is very much in line with the goals of another 'slow movement': CittàSlow (slow city) (Miele, 2008). Frequenting locally-owned and sustainable businesses will also help to keep tourist dollars within the community (Conway & Timms, 2010; Hall, 2006).

A certain mindset can be helpful for those who wish to embrace the concept of slow tourism. With regard to transport, it is essential to perceive the *journey* as being an important (and hopefully enjoyable) part of the experience itself (Dickinson & Lumsdon, 2010; Verbeek & Bargeman, 2008). Also helpful is having a sense of adventure, embracing uncertainty and flexibility. For example, taking advantage of unexpected opportunities, meandering through communities, and accepting the transport delays or changes that may accompany the often increased connections necessary to reach a destination (e.g. bus–train–bus) (Verbeek & Bargeman, 2008). There are also certain actions that will aid in the progression towards slower tourism. Most important is consumer reflexivity, which involves making conscious and responsible choices, particularly with regard to transport (Germann Molz, 2009). It can be helpful for slow tourists to consider transport options before making a final decision about destination choice (Dickinson *et al.*, 2010). For example, some destinations are difficult to reach without using an airplane or private vehicle, which may be an important consideration for slow tourists.

As previously discussed, engagement with local communities is an integral aspect of the slow tourism experience. Nevertheless, immersion and participation may be difficult, as tourists are often perceived as outsiders by local people. Opportunities to participate in locals' lives are provided by programmes such as Willing Workers on Organic Farms, Help Exchange and Work Away, which allow tourists to live with and work alongside their 'hosts'.

Willing Workers on Organic Farms (WWOOF)

Willing Workers on Organic Farms[3] is a form of tourism that is highly conducive to a slower approach to travel. Travellers, or 'WWOOFers', exchange four to six hours of their labour for meals and accommodation at a host's property. These properties are not all farms; some are even in urban areas, although all hosts do some type of organic growing and/or animal-raising. The type of work WWOOFers are asked to do spans a wide range, however, the majority must take place outdoors. It includes, for example, weeding, harvesting, tending to animals and property maintenance.

Although the focus of the research discussed below is Australia, independent WWOOF organisations exist in 49 countries. WWOOF United Kingdom and The WWOOF Association also produce lists of hosts in 67 countries that do not have their own organisations (The WWOOF Association, n.d.; WWOOF United Kingdom, n.d.). Travellers who wish to join WWOOF Australia pay a membership fee (currently AU\$65) to receive a 'WWOOF book' and volunteer accident insurance, both of which are valid for one year. The book contains a description of each host and property, as well as their contact details. Prospective WWOOFers call, email, and/or text hosts they are interested in visiting. If a host is willing and able to accept the applicant, the two parties make further arrangements (e.g. time of arrival, length of stay).

Methods

The data discussed below were collected during five months of fieldwork in 2009 and 2010, as well as through two online surveys. In total, 40 travellers were interviewed in the field, 104 participated in the pre-experience online survey, and 129 answered the post-experience survey. They are independent samples, although there is a core set of questions common to all three groups (e.g. age, country of origin). When appropriate, this common data will be presented in the following discussion as a combination of all three samples. Data that come from only one or two of the samples will be clearly identified.

Potential respondents were approached in a variety of ways. First, for the fieldwork portion of the research, five regions were chosen along the east coast of Australia (Queensland, New South Wales and Victoria). Due to time limitations, it was necessary to restrict the study to only a portion of the country. These states were chosen because they have the highest concentration of both WWOOF sites and tourists and are adjacent to one another. The regional boundaries are based upon categories used in the WWOOF book, although some of them were combined.[4] The choice of region(s) within each state was largely based on WWOOF site concentration and a desire for geographical dispersion. The latter is important in order to capture regional differences, and the former so as many sites were able to be visited as possible within the time available. In addition, the four largest cities[5] were avoided as the researcher aimed to capture the experiences of participants in more rural areas. The proportion of sites in each state's sampled regions approximates that of each state's total population of sites. For example, 40.4% of the properties in the *sample* are in New South Wales, while that state contains 41.4% of the *total* number of host properties. Quota sampling was employed in choosing the number of properties to visit per state and per region. In total, 30 properties were visited in Queensland, 30 in New South Wales and 15 in Victoria.

Within the five regions, 123 properties were randomly selected. Of those properties, 75 were visited (61%), 9 hosts were willing to participate but had scheduling difficulties (7.3%), 23 declined the request (18.7%), the researcher was not able to contact 14 hosts (11.4%), and 2 properties were deemed too far to visit (1.6%) (>7 hours drive inland). All WWOOFers who were at the properties at the time of the researcher's visit were asked to participate. Some properties had no WWOOFers present,[6] while others had several. Only five could not be interviewed: four because of a lack of English fluency and one did not have enough time. The researcher also interviewed one WWOOFer in between sites.

To advertise for the online surveys, the WWOOF Australia team agreed to put 4000 flyers in the books (over approximately four months), as well as post a notice on their website's bulletin board. Some of the hosts visited by the researcher also contacted former WWOOFers regarding the post-experience survey. Potential respondents were directed to the Survey Monkey website, which hosted both the pre-experience and post-experience online surveys.

Results and Discussion

There are numerous connections between WWOOFing and slow tourism, spanning all three elements: reduced mobility, a shift in modal choice and engagement with local communities. Given the fact that there are hosts in over 100 countries, short-haul and/or domestic WWOOFing destinations are available to all but the most remote travellers. Domestic tourists accounted for 17% of respondents, and 3% are from the nearby nations of New Zealand and French Polynesia. Most of the WWOOFers, however, are long- and medium-haul tourists from Europe (56%), the Americas (12%) and Asia (11%).[7] About three quarters of those surveyed consider it likely or very likely that they will WWOOF again in the future (post and on-site samples). The WWOOF Australia website lists all of the other independent organisations around the world, which may help travellers to identify short-haul WWOOFing destinations for future travel.

WWOOF Australia does not advertise overseas, and they are beginning to focus more on domestic markets, such as students of horticulture and agriculture (G. Ainsworth, personal interview, 4 March 2010). WWOOFing locally is a particularly good travel option for those who want to gain knowledge that is directly applicable to the climate and region in which they live. Every host has a different passion and area of expertise, and each property also has something unique to offer travellers. Most WWOOFers are thus able to have a diverse experience, as 65% have or plan to visit at least three properties during their travels. The increasing popularity of the programme also results in new hosts joining every year; in Australia alone there are currently over 2300 hosts to choose from (WWOOF Australia, n.d.). The experience will also be very different depending on the season during which the traveller visits. All of these factors are of great benefit to the local and domestic tourist, as they can re-visit the same areas and have substantially different experiences every time.

Having more to do than simply visit the main sites listed in tourist guide-books can encourage people to stay longer in each location, reducing levels of mobility during the trip. WWOOFers do tend to literally slow down their travels, as some plan to visit a property for a week or two but end up staying several months. On-site and post-experience respondents reported an average length of stay at each property of six more days than the expectations of pre-experience respondents (21 vs 15 days). This difference would be greater if the estimate only included the hosts with whom WWOOFers enjoyed staying. For example, some would stay several months with one host, but only a few days at another property because they did not have an enjoyable experience there. One of the WWOOFers interviewed got along so well with the hosts that she was living with them indefinitely. Furthermore, even when a WWOOFer moves on to another host, it can be within the same area. There are local/regional and informal networks of hosts who recommend 'good' WWOOFers to each other. It is easy to stay busy and interested in a small area, as each host/property differs somewhat from each other and offers a wide variety of experiences. Slowing down was an explicit interest for many, as 45% of those interviewed during fieldwork reported 'settling or resting from constant travelling' as an important or very important reason to WWOOF.

Depending on the arrangement with their hosts, WWOOFers generally only have up to a day or two off from work each week to explore the area. The other days are usually quite full with four to six hours of work plus help-ing with cooking and/or cleaning and some relaxation time. This results in much less travel around the destination and more time spent exploring the property, reading, or simply sitting on the veranda enjoying the views and chatting with hosts. Even on days off, nearly 60% of WWOOFers do not have access to a personal vehicle and thus depend on the generosity of their hosts to visit any local sites that are outside of walking or cycling reach (post and on-site samples). Some hosts are happy to be an informal tour guide for their guests, while others only do this for those who put in an extra effort on the property. Whether or not travellers participate in fewer motorised activities also depends on the host. Many do not have motorbikes, all-terrain vehicles, etc. and some hosts who do will not allow WWOOFers to use them because of liability concerns. At other properties, these types of machines are an important part of the work and leisure time.

Most hosts will pick up WWOOFers at the closest bus or train stop, allowing travellers who do not have their own vehicle much easier access to rural areas. This is very important for short- and medium-haul travellers, as it could encourage them to forego travelling by plane or private vehicle alto-gether. It could also help to promote the use of shared surface transportation for travel between destinations. Once in Australia, the main forms of trans-port used by respondents are: buses (61%), personal vehicles (42%), trains (23%), planes (12%) and hitch-hiking (10%) (post and on-site samples).

Although 42% and 12% are fairly high proportions of people using personal vehicles and planes, the usage may be lower in smaller countries and those with faster and more comprehensive train systems. For example, Australian train routes are not well connected as the main ones only branch off from three large southern cities, which can lead to a lot of unnecessary back-tracking (Australian Rail Maps, n.d.). Bus and train passes are also relatively expensive in Australia; it can be cheaper to buy a car when travelling with a companion and for an extended period of time. These are issues which should be addressed by relevant government organisations and private providers.

With an average of two to three weeks at each property, a long trip is an easy undertaking even in a relatively small area. Longer trips are also much easier to afford due to the room and board cost savings which, particularly in developed countries, are quite significant. Of respondents, 89% were travelling for more than two months, and 75% have or plan to spend over $2,000 during the trip (not including airfare to and from their home country). Travellers will rarely spend their entire trip WWOOFing, allowing time for other activities and visiting different sites, with some also undertaking paid work and attending language schools. The lengthier trips still leave plenty of time for WWOOFing, as over 70% of those surveyed have or plan to participate for over a month.

One of the biggest obstacles to slow tourism is how an 'outsider' can *truly engage* with locals, beyond those who work in the tourism industry. WWOOFing provides travellers with an opportunity to live with and work alongside hosts, allowing for greater insight into how they live. Getting to know locals/hosts was an important or very important reason to WWOOF for 82% of respondents (pre and on-site samples). Furthermore, the vast majority (96%) stated that interactions with their hosts were a (very) important part of the overall experience (post and on-site samples). Often hosts will include invitations for visitors to attend dinner parties and other social events to which they have been invited. This does not go unappreciated, as being included in hosts' social activities is (very) important to 70% of respondents (post and on-site samples). WWOOFers are usually treated like 'one of the family' which enables them to experience all aspects of the lifestyle, not just the pleasing ones that put on a good show for normal tourists.

Immersion in the local culture and getting to know the places visited were also (very) important reasons to WWOOF for many respondents (75% and 81%, respectively) (pre and on-site samples). Food is a wonderful way to get to know a culture and region, and eating with their hosts is (very) important to 82% of those surveyed (post and on-site samples). WWOOFers not only have the opportunity to eat local and seasonal food, but are also able to help grow, harvest and prepare it. These travellers become *participants* in these rich experiences instead of simply observers.

Lastly, there are abundant opportunities for exploring local natural resources during the WWOOF experience. It can often not be avoided, as working with the land is generally part of the job requirements. Spending time in rural/natural areas and experiencing rural/farm lifestyles was a (very) important reason to WWOOF for 86% and 83% of those surveyed (pre and on-site samples). During their leisure time, exploration is facilitated by hosts' local knowledge of good spots to visit and safe trails to wander, and at some properties the provision of bicycle and/or horse transport. Particularly in the more rural areas, entertainment often involves enjoying simpler pleasures: walking through the forest, swimming in a creek, fishing, reading, getting to know new companions and enjoying a home-cooked meal. One of the most promising aspects of the WWOOF programme is that it provides the opportunity for suburban- and urban-based people to develop a deeper connection to the land and awareness of how food is produced. About half of those who participated in this research have spent the majority of their lives in a medium to very large city (100,000+ inhabitants).

Although WWOOFing may not interest everyone, the only participation requirements for travellers are the ability and willingness to work four to six hours per day. The nature of the work also necessitates 'at least average physical fitness' (WWOOF Australia, 2009: 10). Nevertheless, different work arrangements may be made with hosts. For example, an interviewed host has a wheelchair-accessible house and is happy to accept persons with disabilities. The WWOOF experience is open to all backgrounds, skill levels, and languages spoken, as well as most ages. The maximum age to join WWOOF Australia is 80 and the minimum is 18; however, children may accompany adults to properties if prior arrangements are made with hosts.

The eldest WWOOFer surveyed was 66 years of age, although over three quarters of respondents were less than 30 years old. The age distribution is much more even amongst the Australian respondents: 38% under 29, 30% between 30 and 39, and 33% over 40. At least part of the explanation for this may be due to the popularity of the Australian working holiday visa, which is only available for people less than 31 years of age. Although travellers do not need this visa in order to WWOOF in Australia, it allows for a 12-month length of stay with the opportunity to undertake paid work. Longer trip lengths are an important option for slow tourists who take fewer overall trips, use slower modes of transport and seek deeper engagement with local communities. Given the longer trip lengths and unstable global economic situation, the opportunity to earn money can be very attractive. Working holidaymakers from some countries can also apply for a second year-long visa if they have undertaken specified work for three months in regional Australia. As many WWOOF properties qualify for this scheme, a visa extension was a (very) important reason to WWOOF for 30% of those surveyed (pre and on-site samples). The immigration departments of some countries (e.g. New Zealand) do require a work visa in order to WWOOF, which is age discriminatory and a significant hurdle to the growth of the programme.

Conclusion

WWOOFing is an important option for the growing slow tourism movement, even if it does not suit everyone. For those not explicitly interested in slow travel, WWOOFing can nevertheless facilitate and encourage a slower and more sustainable approach to tourism. Some WWOOFing activities, such as assisting hosts with revegetation projects, may 'offset' environmental damage done during other parts of the trip. Nearly 60% of respondents stated that 'giving back to the country/land' was a (very) important reason for them to WWOOF (pre and on-site samples).

WWOOFing could also play a small role in the progression towards more sustainable tourism choices for future trips. The vast majority (89%) of post-experience and on-site respondents feel more committed to the environment as a result of their WWOOF experiences. This will hopefully encourage more conscientious tourism decisions in the future, provided they are aware of the detrimental environmental impacts associated with tourism. This environmental commitment may also lead to more sustainable choices beyond the tourism realm. A (very) important reason to WWOOF for three-quarters of those surveyed was 'experiencing a more sustainable lifestyle' (pre and on-site samples). With around 13,000 books sold by WWOOF Australia alone in 2009, the reach of the programme is quite far (T. Wilson-Brown, personal interview, 5 March 2010).

Even if slow tourism remains a niche area that is only embraced by a small portion of tourists, it is absolutely essential that some elements of this practice have widespread adoption in light of sustainability concerns. This is necessary in order to avoid serious, irreparable harm to the earth and its inhabitants through global warming. Tourism's current (and projected future) contribution to global warming must be reduced through a focus on low-carbon travel. For the foreseeable future, technological achievements will be nowhere near sufficient to tackle the problem. First and foremost, a shift towards shared, surface transport and away from planes, cruise ships and cars is vital. Taking fewer trips and travelling less within each destination is also important (except for when using self- or animal-propelled transport). Lastly, people need to start focusing more on the interesting destinations available closer to home. The methods through which these lofty behavioural changes might be achieved are beyond the scope of this discussion, although the increased levels of environmental commitment reported by WWOOFers could lead to more conscientious decisions in the future. The solution will lie not only with consumers, but also in concerted efforts by governments and tourism/transport providers.

Interestingly, many slow travel proponents focus solely on experiences at the destination level, excluding transport to and from the destination(s) as an important consideration for slow travellers (Dickinson & Lumsdon, 2010). This is a significant oversight, as travel to, from and between destinations is an important aspect of all tourism experiences. Interest lies not only with the environmental impact of carbon-intensive modes of transport, but also with the potential to enjoy slower forms of travel as part of the overall tourist experience. Environmental sustainability has always been a core element of the original slow movement – Slow Food. Their vision is, 'a world in which all people can access and enjoy food that is good for them, good for those who grow it and *good for the planet*' (Slow Food International, n.d., emphasis added). WWOOFing also has strong ties to this movement as it provides the opportunity for people to participate in local food growing, harvest and preparation. A greater awareness of how food is produced and the impacts involved in this process are essential to environmental sustainability, as 'how and what we eat determines to a great extent the use we make of the world – and what is to become of it' (Pollan, 2006: 11).

Notes

(1) The radiative forcing figures used in this chapter include the maximum contribution of contrail-induced cirrus clouds.
(2) Hitch-hiking may be an exception, however, as it takes advantage of and leverages off a pre-existing trip.
(3) Some of the individual organisations have re-named their programme 'World Wide Opportunities on Organic Farms'.
(4) Far North Queensland (from Cape Tribulation south to Silkwood), South East Queensland (from Gladstone south to Landsborough), North East New South Wales (from the Queensland border south to Woodburn), Central East New South Wales (from Kempsey south to Gosford), and Eastern Victoria (Gippsland and the north east).
(5) Brisbane, Canberra, Melbourne and Sydney.
(6) Hosts were also interviewed, although these results are reported elsewhere.
(7) As the surveys were only available in English, those from English-speaking countries may be overrepresented in this sample.

References

Australian Rail Maps (n.d.) Australian rail map – Online document: http://www.railmaps.com.au/
Conway, D. and Timms, B.F. (2010) Re-branding alternative tourism in the Caribbean: The case for 'slow tourism'. *Tourism and Hospitality Research* 10 (4), 329–344.
Dickinson, J. and Lumsdon, L. (2010) *Slow Travel and Tourism*. London: Earthscan.
Dickinson, J.E., Robbins, D. and Lumsdon, L. (2010) Holiday travel discourses and climate change. *Journal of Transport Geography* 18 (3), 482–489.
Germann Molz, J. (2009) Representing pace in tourism mobilities: Staycations, slow travel and The Amazing Race. *Journal of Tourism and Cultural Change* 7 (4), 270–286.

Gössling, S. (2002) Global environmental consequences of tourism. *Global Environmental Change* 12 (4), 283–302.

Gössling, S., Hall, C.M., Peeters, P. and Scott, D. (2010) The future of tourism: Can tourism growth and climate policy be reconciled? A climate change mitigation perspective. *Tourism Recreation Research* 35 (2), 119–130.

Hall, C.M. (2006) Introduction: Culinary tourism and regional development: From slow food to slow tourism? *Tourism Review International* 9 (4), 303–305.

Hall, C.M. (2009) Degrowing tourism: Décroissance, sustainable consumption and steady-state tourism. *Anatolia: An International Journal of Tourism and Hospitality Research* 20 (1), 46–61.

Knight, J. (2010) The ready-to-view wild monkey: The convenience principle in Japanese wildlife tourism. *Annals of Tourism Research* 37 (3), 744–762.

Matos, R. (2004) Can slow tourism bring new life to alpine regions? In K. Weiermair and C. Mathies (eds) *The Tourism and Leisure Industry: Shaping the Future* (pp. 93–103). Binghamton, NY: Haworth Hospitality.

Miele, M. (2008) CittàSlow: Producing slowness against the fast life. *Space and Polity* 12 (1), 135–156.

Pollan, M. (2006) *The Omnivore's Dilemma: A Natural History of Four Meals*. New York: The Penguin Press.

Scott, D., Amelung, B., Becken, S., Ceron, J-P., Dubois, G., Gössling, S., Peeters, P. and Simpson, M.C. (2008) Technical report. In *Climate Change and Tourism – Responding to Global Challenges* (pp. 23–250). Madrid: World Tourism Organization.

Slow Food International (n.d.) Our philosophy – Online document: http://www.slowfood.com/international/2/our-philosophy

The WWOOF Association (n.d.) List of independent hosts – Online document: http://www.wwoofinternational.org/independents/

Verbeek, D. and Bargeman, B. (2008) The sustainability chain is as strong as its weakest links: A case on sustainable tourism mobility in the Alpine region. Paper presented at the *Conference of the Sustainable Consumption Research Exchange* (SCORE!) Network, Brussels, Belgium, 10–11 March.

Willing Workers on Organic Farms (WWOOF) Australia (2009) *The Australian WWOOF Book*. W Tree, Victoria: WWOOF Pty Ltd.

Willing Workers on Organic Farms (WWOOF) Australia (n.d.) Home page – Online document: http://www.wwoof.com.au/

World Tourism Organization (UNWTO) (2010) Tourism highlights – Online document: http://www.unwto.org/facts/menu.html

World Wide Opportunities on Organic Farms (WWOOF) United Kingdom (n.d.) WWOOF independents (UK) – Online document: http://www.wwoof.org/independents.asp.

Part 3

Slow Mobilities

8 Gendered Cultures of Slow Travel: Women's Cycle Touring as an Alternative Hedonism

Simone Fullagar

I think it's addictive. I say to my kids, I don't do drugs, I don't do alcohol, I don't gamble: I ride a bike . . . . The only holidays I have now are cycling holidays. (Sally, 57 years)

In this opening quotation Sally evokes the compelling pleasures of slow travel and the desire for cycling that motivates many women to begin a different kind of journey. With the rise of the slow movement there has been surprisingly little attention paid to the gendered cultures that inform and are informed by discourses of fast or slow living. In contrast, the mobilities literature has begun to map out some discursive parameters that give shape to gendered cultures of movement (and stasis), embodied and reflexive performances of identity and different perceptions of risky and pleasurable practices (Cresswell, 2010; Sheller & Urry, 2006; Uteng & Cresswell, 2008).

In this chapter I bring together work on gendered mobilities and cultural studies of consumption to offer a feminist perspective on slow travel through an analysis of my own and other women's experiences of cycle touring. I draw upon ethnographic research and interviews with 17 women cyclists (aged 25–75 years) who I accompanied on a nine-day ride over 600 km with over 1000 riders on an organised tour with Bicycle Queensland in Australia, 2010. Being careful not to essentialise women's (or men's) experiences as homogenous I explore how women, as gendered subjects, articulate their desire for cycling as a form of slow travel. As with all domains of life, slow tourism is not free of gender and this chapter offers to shed light on the multiplicity of slow ways of moving and engaging with the world through the specificity of the body, and through the relations between bodies that are also circumscribed by power differences (Bonham & Koth, 2010; Furness, 2010; Pritchard & Morgan, 2000).

While sustainable tourism continues to be a highly contested area in theory and industry practice, the emergence of a slow travel ethos brings with it possibilities for moving, knowing and relating differently (Hall, 2009). Taking inspiration from the slow food movement, the parameters of slow travel and tourism have been defined in relation to the meaningful experience of the journey that embraces a slower temporality, concern about environmental impact and local diversity (Dickinson & Lumsdon, 2010). Yet, most literature has focussed more attention on environmental concerns and temporality, than the convivial or social experience of the journey (Dickinson *et al.*, 2010; Germann Molz, 2009). Mass cycle tourism events constitute a particular form of slow travel, although most events do not self-consciously identify with the slow movement in their promotion and not all participants in this research made an overt connection with the slow ethos. However, as I shall argue in this chapter, the meanings the women articulated about the cycling journey in many ways reflected key dimensions that define the slow movement (Germov *et al.*, 2011; Parkins & Craig, 2006; Tam, 2008). For example, cycling figures as an embodied journey that creates a different temporal relationship with local places and cultures; the collective context of a mass tour offers a sense of conviviality and participants negotiate their desire for mobility and greener travel experiences.

As a form of sustainable travel, cycling can be low impact, however, there are many ways of undertaking cycling journeys across the spectrum from light to dark green (Dickinson & Lumsdon, 2010). Mass cycling as slow tourism is, in one way, a low carbon form of mobility that contributes to greater sustainability through its influence on everyday lifestyle practices and directly in terms of tourism choice. Yet, there is the perennial issue of the mode of transport chosen to begin the cycling journey. My focus on the Cycle Queensland tour event also raises questions about the carbon footprint created by the many large trucks that are required to move the camp, luggage and staff to a new location each day. This is offset to some degree by 1000+ individual riders choosing not to use cars for their week-long holiday, although a number may have driven or flown to the start of the ride. Bicycle Queensland actively promotes the use of coaches and trains as more sustainable alternatives for riders and there are identified social and economic benefits created for the many (disadvantaged) rural communities (see Bicycle Queensland, 2009). More analysis is needed of the impact of the tour event and the benefits arising from the uptake of cycling in everyday lifestyles as well as holiday choices. However, what this chapter offers is an exploration of the slow meaning of mass cycling touring, which can in turn inform debates and tourism practices that are focussed on the issue of sustainable provision and consumption.

Debates about green lifestyles and sustainable travel can tend to be infused with a kind of environmental moralism that is often defined against (or ignores) the desire for pleasurable experiences of places, mobility and connection with others. For example, a recent article by Jackman (2009: 20) in *The Weekend Australian Magazine* stated that 'your average green activist still seems like a humourless and hectoring pedant who will not quit their campaign to save the planet until they have destroyed any scrap of enjoyment we gain from living here in the first place'. Slow tourism and travel potentially offer a way to think through the relationship that connects environmental concern and pleasurable experiences that can sustain lifestyle change and transform highly consumerist practices. It is from this perspective that I draw upon Kate Soper's (2008) notion of 'alternative hedonism' as she evokes the ethical relations between pleasure in different consumption practices and the rising disenchantment with consumerism. Soper (2008: 5) argues that Euro-American notions of the 'good life' rest on a privileged ideal of affluent consumerism and are only likely to be countered with appealing ideas about what it means to enjoy a high standard of living. Rather than simply offer an environmental critique of hyperconsumption that simply dismisses consumer desires, Soper suggests that the emergence of more sustainable practices will be dependent upon new ways of thinking about enjoyment and pleasure in everyday life. In this sense, the accusation that the slow movement is simply a middle-class indulgence is an argument that misses an important point: sustainable leisure and travel options need to become desirable options as part of the rise of the ethical consumer. Rather than dismiss the tourist's desire for slow experiences, products and services, how can we think about slow travel as part of the emergence of a 'new hedonist imaginary' (Soper, 2008: 571) that embraces an anti-consumerist ethic and the politics of consuming differently? Slow travel can also be thought about as a gendered experience of mobility that is produced within the sphere of 'life politics' that shapes women's leisure consumption and identity formation in advanced economies (Rojek, 2010; Rose, 1999).

Researching Gender and Cycling

> ... even if it is the same road, [it is] a different experience. (Kim, 60 years)

Like other traditionally male sporting cultures, much of the research on cycling as a recreation, competitive sport or touring experience has assumed the masculine subject as the norm (see Dickinson & Lumsdon, 2010; Dickinson & Robbins, 2009; Hodgson, 2007; Lamont & Causley, 2010; O'Connor & Brown, 2007; Ritchie *et al.*, 2010; Spinney, 2009). A number of historical (Simpson, 2001) and contemporary studies (Hanson, 2010; Womack & Suyemoto, 2010) have identified the freedom and empowerment that cycling creates for women via increased independent mobility (away from the gendered constraints of home). There has also been a focus on understanding the gendered barriers to cycling and how women's participation can be increased to address inequities, target women as cycling consumers, encourage green transport and promote healthy lifestyles (Emond *et al.*, 2009; Garrard, 2003; Garrard *et al.*, 2008).

Although Australia is far behind most European countries in terms of participation, infrastructure and cycling inclusive culture, cycling has grown in popularity. As the fourth most popular physical activity, 6.5% of Australians over 15 years cycled for recreation or sport at some time in 2009–2010. Yet, if we look more closely at the gender differences we see that there are still far fewer women cycling in their leisure time (men 8.2% and women 4.9%) (Australian Bureau of Statistics, 2010). Commuting by bike is on the increase in many cities and cycle tourism has begun to take off with rail trail development and a range of small and mass cycle tour events that are held in many states (Ritchie *et al.*, 2010). There are many forms of cycling and ways of moving, and not all of them embrace a slow ethos. Indeed the fast pace, expensive bikes and competition are often implicit in the comment that cycling is on the rise as the 'new golf' for businessmen. A number of women enjoy this kind of cycling, too, and I do not wish to imply that this is morally questionable, rather my interest is more specific to cultures of slow mobility and slow tourism.

At a time when a number of traditional sport clubs are experiencing declining membership, state-based, not-for-profit bicycle associations have had rapid growth. For example, Bicycle Queensland, who organise the Cycle Queensland tour that is the focus of this research, doubled its membership over the last five years to reach over 10,000 members (Bicycle Queensland, 2009). Yet, women do not constitute 50% of members or 50% of riders on the 1000 plus Cycle Queensland tour, and hence the aim of this research was to understand more about the gendered context of cycling as a sustainable form of everyday travel and tourism. Given the lack of research into women's experiences of large organised cycle tourism events, like Cycle Queensland, I wanted to consider both conceptual and practical issues relating to the gendered culture of cycling to encourage different ways of thinking about what might encourage more women to participate. With the support of Bicycle Queensland and some university funding I was able to interview 17 riders (from 40 women who responded to my email invitation) aged 27–71 years. The majority of the women were from an Anglo-Celtic background and 11 were experienced cycle tourists while six were first-time riders on this kind of event. They came from mixed social backgrounds, with seven women residing in regional or rural areas in different states of Australia (apart from one New Zealander) and 10 residing in cities or large urban areas. Four women attended by themselves (often meeting up with friends or an organised group), another four came with their own self-organised group (often a bicycle user group or fund raising group of friends), five came with their partners, three came with a female friend and one woman was riding with her partner and three children for the ninth year in a row. I completed interviews at the campsite on the rest day or once riders arrived at camp for the afternoon. Being part of the ride was important in terms of the ethnographic journey, the collective context of the event and to be able to share something of that experience with the participants.

Using a reflexive approach to analysis I identified several key themes across all the responses and interpreted these in relation to feminist post-structuralist work on mobilities, consumption and tourism (Alvesson & Skoldberg, 2000). The key themes that I will focus on in this chapter include women's narratives of the 'slow journey' (often defined in relation to a masculine culture of speed), the sense of becoming part of a 'convivial community' and the relationship between 'hedonic identity and sustainable mobility'. I argue that these dimensions of cycle touring embody an alternative hedonism that Soper (2008) associates with the rise of non-consumerist practices and environmental critique. Next I want to offer an ethnographic glimpse into the liminal journey of the ride by bringing slow mobilities into the meaning-making process of tourism and event research (see also Holloway *et al.*, 2010; Watts & Urry, 2008). As Spinney (2009) argues, cycling has conventionally been 'known' through rational and linear assumptions about movement rather than through methods that value the ephemeral, the fleeting and the sensual.

An Ethnography of a Cycle Tour Event

As a cycle tourist I joined over 1000 riders on several annual journeys through rural Queensland to enjoy the pleasures of slow travel and safety in numbers on the (often life threatening) road. On day one of Cycle Queensland we move out en mass from the town centre as a swarm of bikes and bodies streaming for miles, up and down country roads for about 80 kilometres each day. 'Cycle away from the everyday' is the event strap line on the marketing material that accompanies waving images of male and female cyclists of all shapes and sizes. We are urged to escape our sedentary existence and embrace the embodied challenge of finding our own pace along with the sociability of moving together. For nine days we become a nomadic community, of which two thirds are men and one third women, bound by affective relations that connect and separate us; alternately we love and loath the rolling terrain, heat or rain and the rhythm of organised ride–camp life. The embodied mass is also fractured into differences marked by corporate jerseys, a competitive or leisurely ethos, bike brand and type, family or friendship group, staff, volunteer or rider, different ages, sexualities, cultures and genders. Women of all ages and abilities come on the ride in couples with male or female partners, sometimes alone or with groups of friends, cycling clubs or family. Every day cycling is a search for the slow rhythm with its alternate states of immersion in the immediacy of places and the focused intensity of physical challenge. It is awakening movement that counters the tendency to forget how to live in the present with our car-centred, work-oriented lifestyles. At times I overtake labouring, or leisurely, riders ploughing the headwind, we chat briefly about the day and offer encouragement. In turn I am overtaken by those, often years older than I, seeking the momentary thrill of the downhill.

Each evening the community of cyclists comes together in the middle of our tent city to dine under the stars and listen to the ride briefing about weather, road conditions and local knowledge. To test our tired bodies further we are offered a tempting array of choices, with live music, an outdoor cafe and movies, as well as local towns to explore. The culture of the ride is convivial, inviting social exchange between strangers, from fixing flats to yarns about previous rides or future journeys. Daily adventures on the ride are shared in shower queues or breakfast line ups, ideas about bike assembly or riding tips are swapped with new friends. Some relearn the lost art of relaxation after physically demanding effort and retrace the connection with people met on previous rides. The cycling tour community is a fleeting formation in time and space, offering a collective escape from the work-consume treadmill and a chance to remember the shared pleasures of slowing down, living simply.

Slow Journeys: It is Not a Race

> I like the idea of the journey, rather than an event – I don't mind events, but the idea of travelling as part of cycling was really good for me ... it's the pleasure of applying yourself to something to improve without having to then put yourself against some kind of stopwatch that takes the pleasure out of it. (Kim, 60 years)

Standing in the registration queue the day before the ride begins I strike up a conversation with the older woman behind me who is keen to get her number and reassemble her bike that is boxed in multiple pieces. 'Molly' (56 years) happens to be one of the respondents that I am going to interview during the ride. Having established this fact within several minutes we discuss our delight in the journey ahead; no cooking or cleaning, long days pedalling through the countryside and some daunting steep hills to test our fortitude. We catch up for an interview after several days of riding when the rain starts and does not stop for several days, just to test our tolerance for mud. Molly is not phased at all by the rain and talks about how she has planned well ahead to do the ride with her friend because her husband does not come anymore (his snoring woke the campsite and he has a different cycling 'attitude'). Like many of the women I interviewed, Molly enjoyed the embodied challenge, health and environmental benefits of a cycling holiday (she is very fit), that is also safe and well organised, 'I really like the feeling of working out big time during the day while seeing beautiful places and coming home tired, exhausted and hungry ... I enjoy the fact that it doesn't take up any petrol to ride a bike.' She was given a bike at age 45 and along with her regular rides, alone and with friends during the week, she has been on several small and large cycle tours in the last 10 years. The cycle tour keeps her motivated and she looks forward to being,

> away from the everyday. I find that immediately you are out on your own, or with friends, I notice the birds, I notice the spider webs with the dew on them ... I never wear anything in my ears ... I just really enjoy having a lung full of air and the freedom ... seeing the countryside, different places that you go to slowly ... the trouble people go to when we all arrive (in small towns), people waving and morning teas. (Molly, 56 years)

Molly invokes a slow ethos in responding to my question about what it is she enjoys about the cycle tour journey. In a similar way, experienced cycle tourist Dale (45 years) talks about mobility as simplicity, 'travelling by bike is really my ideal way of feeling like you're actually travelling through an area not just going from A and then arriving at B and hopping out. You actually experience the whole journey. I just love the freedom and the simplicity'. The journey is described through the language of the body and sensory engagement. Jean (68 years) talked about the immediacy of moving through the time-space of cycling as something that characterises the slow journey:

> Well it's probably the only time I actually can get in the moment. Like you really have to be in the moment when you're cycling. You have to be aware, you're listening for magpies and noises. You're watching, you're scanning for brake lights going on thinking somebody's stopping or going to turn at you. You're concentrating. So I can get out of my head I guess. I'm a thinker. So I get out of my head when I'm cycling. I just cycle along and I compose songs and think great thoughts. ... You get on your bike and your legs have got muscles and it just feels so good .... The pleasure in the body being stronger is really exciting yes. So that's probably it I suppose and the wind in your face. The fact that you're travelling. I go to sleep in cars; I really don't like driving in cars much. So you feel very alive on a bike because you're using all your senses.

Other women spoke about slow mobility in spiritual terms, as Joan (60 years) said: 'I do a lot of communing with God when I'm out on my bike. I love ... seeing all the animals, you saw those cattle lined up at the fence .... I love seeing the countryside and looking at the houses and I wonder what those people do and it's "me" time'. Women's experiences of the slow journey were often articulated in relation to gender differences and contrasted against a masculine culture of speed. Molly (56 years) defined her enjoyment of the slow journey against her husband's focus on speed, distance and the point of arrival:

> He has a totally different attitude to cycling. He gets on his bike and he gets a big sweat and he wants to get there, whereas I'm happy to have morning tea and to chat, to take it easy and to look at the view. Not just to see what time I leave and how fast I go and what's my average speed, all that kind of stuff!

Kim (60 years) also reflected on the masculine desire for speed as a display of strength to others 'that goes to power – it goes to strength and power which is more important to them than health'. The counter discourse of slow was evident in many women's comments about how important it was that the cycle tour enabled a leisurely journey rather than normalising a competitive ethos. Sally (57 years) spoke about this when she said, 'I wouldn't even call it a sport, [perhaps] a recreational activity, because a sport has that inference that you're racing or you're competing … it's more the enjoyment of morning tea, (riding) chatting with people and lunch, (riding) chatting with people.'

Many women identified the value of riding at one's own pace, the lack of competitiveness that might show women up as less able than men and the feeling of safety that the event organisers created. Joan talked about the importance of riding at her pace:

> the tour is very, very good for that sort of thing. Nobody knows when you leave or when you get in. Because I mean a lot of the blokes have been in town for hours at the pub so then they straggle in the afternoon at the same time you're coming in. So there's no competitiveness about it at all. You don't feel pushed. (Joan, 60 years)

For many women, the slow journey was defined against masculine notions of speed and competition, as well as less sustainable transport such as cars.

Convivial Communities

> It's a lovely community. I mean look at them. We've been in the rain for all this time and they're still happy. They're still laughing and joking … . It is unique yes. We've got away from the materialistic side of the world and they're just concentrating on each other and what's important … . Yes in nine days. It's almost like a retreat isn't it? A retreat from life. (Jean, 68 years)

In addition to the slow pleasures of cycling, the next major theme was the sense of a convivial community that was created during the tour. The pleasure of sociability, of meeting new and familiar people, and sharing the journey with others was a strong focus of the nine-day tour. In Soper's (2008: 580) terms these social relationships provide the basis for an alternative hedonism, as Jean implies in her comment about 'getting away from the materialistic side of the world'. Sally (57 years), as a solo rider and volunteer, spoke about the feeling of community that was created by 'the camaraderie of the cyclists. I enjoy the bike riding, I enjoy being out in the countryside but the main thing is the camaraderie of the cyclists that you can sit down to anyone at lunchtime, breakfast, in the queue to the loo and chat away. To me, that's how society should be and I really appreciate that.' As a source of affinity that generated the mobile community, cycling was identified by many women as an alternative marker of identity that enabled more egalitarian interactions. Kim (60 years) spoke about this by saying that:

> What I find is really interesting is that you have no idea about people apart from that they're cyclists when you meet them here. You don't have the trappings of normal life, so somebody could be a corporate executive I guess, or I don't know, clear drains or something. You've no idea when they're on their bike in Lycra, so it's a leveller really ...

For some of the younger participants the social aspect of the ride provided an opportunity to meet potential partners who had similar interests (several long distance romances occurred after the tour) and to also meet other cyclists from their home towns. For Pam (43 years), who had been on all eight annual rides with her three boys (the eldest is now 14) and husband, the tour was a significant site of family leisure connections: 'there're a lot of people that we only know from the ride, that we only meet every year and they watch our kids grow up and each year come and go, wow look where they are now'. Describing herself as shy, Pam undertook the first ride to meet other people and many years later finds herself organising many social rides during the year and providing advice to encourage other mothers to take their families.

All participants spoke about the importance of the inclusive culture of the tour for their enjoyment and how the organisers facilitated this convivial atmosphere through their communication about respect, helping each other out and being patient on the road. The liminal nature of the tour community created a context where social norms relaxed and interaction with a diversity of people was expected as part of the slow experience. For women this sense of convivial community was significant in enabling social relationships (especially for solo riders) that were based on their identity as cyclists (rather than as mothers, partners or workers). The issue of gender identity was also central to how women created meaning about the pleasurable challenge of the slow journey and responded to concerns about environmental sustainability.

Hedonic Identities and Sustainable Mobility

For a number of women, cycling also embodied a slow travel ethos about sustainability that was defined against speed and car-based mobility. Jean (68 years) talked about the Australian

> love affair with the car. All our cars drive much too fast for their own good and the good of everybody else. Most of the accidents that happen these days aren't accidents. I can't understand why we've got such angst against cyclists. Why haven't we got angst against cars? I mean cars kill people ... cycling puts you back in the community. A car takes you out of the community.

Most participants talked about a strong interrelationship between their cycle tourism choices and cycling at home for recreation and/or commuting. Some were inspired by their tourism experiences to cycle more often as part of a greener, healthier lifestyle, while others found that their everyday riding generated a desire to try cycling holidays instead of motorised travel. Yet, there was also an awareness of the contradictions of travelling by plane or car to participate in the cycle tour, as Kim stated about her trip from New Zealand, 'I think environmentally obviously it's really good, except that I was thinking last night I've just punched holes in the stratosphere to get here so I can ride a bike; be a model of carbon neutral or something.' Only a couple of participants did not consider the environmental impacts of cycling to be important in their decision-making, rather they viewed any benefit as a by-product. Lorna talked about this: 'doesn't rate that highly for me. Because, I mean at home I would not be a person that would choose a bike over a car because of environmental issues. I'm just way too busy'. Frequently women talked about the effect of having to juggle multiple gender responsibilities (caring for others, work) on their time. In this sense cycling as slow travel is not always motivated by environmental concern, but it can generate pro-environmental behaviour through tourism choices that are not necessarily intentional (nor does it necessitate a 'green' identity).

In contrast, many women identified themselves as strong advocates of cycling as an empowering experience, where pleasure could be derived from challenging oneself and engaging with the world differently. Kim (60 years) spoke about this, 'I feel very powerful on the bike; I feel very good on the bike.' What is interesting about the desire to engage in sustainable travel was the connection to what I have termed a 'hedonic identity', where women fuse the meaning of pleasurable challenge in their lives with either an implicit or explicit environmental awareness as 'cyclists'. For example, Joan initially talked about how the enjoyment of cycling had become part of her identity as an independent older woman, after buying a bike post-divorce when she was 57. She said 'you feel so good. Your whole body feels (good), you move better'. The Cycle Queensland trip was her first tour and she decided to go to 'give myself a 60th birthday present and this is it. So here I am sitting in the rain at Mount Morgan, loving it'. Joan valued the environmental bene-fits, but like many other women in the research she talked about first having to overcome gender constraints that affected her motivation and capacity to cycle. Jean talked about doing the ride after her husband passed away and how her confidence had been restored after this loss, 'I actually questioned whether I could do things. I've never, ever questioned that I couldn't do any-thing. This has sort of made me see that I can, I can do anything again … it does give you a bit of an identity especially at 68.'

Amongst participants there was a strong gender discourse about cycling as a new source of slow pleasure and identity for women, as they desired an alternative to the gendered responsibilities of home (often tied to con-sumerism). Cycle touring offered the freedom of mobility, social connec-tion and overcoming the gendered limits of self. Lucy (56 years) responded to the question about what she enjoyed most about the tour with 'I've got nine days without having to think of cooking.' As a mother of nine chil-dren in a regional town, Jean (68 years) reflected on her life prior to taking up cycle touring, 'my life was very directed for 31 years and I've been a mother and a wife for such a long time I thought well it's probably about time I found out where I was again, so the 2002 ride was a ride back to "me"'. This notion of cycling as a source of independent identity for women was evident throughout the interviews and illustrates the gendered mean-ing that women attribute to slow travel experiences in relation to the rec-ognition afforded by others. As Kate (46 years) said, it gave her a sense of 'self worth, I think that helps a lot, because you achieve something on these rides. You go back home and you tell people where you've ridden, they go what? You did what? I'm a teacher and the kids say … and they go "wow"'.

A number of women identified how the embodied pleasure and challenge of riding was interconnected with the slow experience of place during the ride and in their communities at home. For Jean (68 years) the pleasure of cycling in her regional area was key to promoting sustainable transport, healthy com-munities and safer roads, 'I'm very passionate about sustainable transport, obesity, and all that sort of thing. Because my children all rode bikes but I've seen the car parks at school just get fuller [with] cars and less [with] bikes. So I'm very passionate about that area.' Pam (43 years) also talked about the effects of the cycle tour on her family's identity and how cycling has been incorporated into their commuting and recreation. She said that is has

been a change of lifestyle I suppose in a lot of ways. I ride to work every day now and things like that, so I have that level of fitness all the time. People say to me when they hear that we're doing the ride, they go, oh you must be so fit – oh that's amazing. But I don't feel any different because that's what I do all the time.

For some women cycle tourism was clearly a form of alternative hedonism, in Soper's (2008) sense, of being motivated by an anti-consumerist intent and desiring other forms of sustainable mobility in everyday life and when travelling. In this way the research contributes to the growing body of knowledge that articulates a gendered understanding of adventurous forms of travel and leisure where the masculine is not presumed as the norm (Hanson, 2010; Little & Wilson, 2005).

Concluding Remarks

Thinking about the pleasurable experience of slow mobility is important for considering how sustainable forms of travel can be promoted to women given the gender disparity in cycling participation rates. While the environmental benefits were mentioned by nearly all participants as part of their cycling motivation, more often women spoke about their desire to be fit, active and healthy, connect with like-minded cyclists as part of a convivial community and experience the immediacy of new places and people. It was the hedonic dimension of slow mobility that motivated most participants through their identity as 'cyclists' to undertake a tour and to cycle for recreation or transport. The findings in this study illustrate how slow travel and the meaning of cycling mobility are gendered in particular ways for women (Uteng & Cresswell, 2008). Without essentialising gender differences I argue that there are significant implications for tourism operators (cycling and beyond) in considering how they develop, promote and support gender inclusive journeys that respond to the multiplicity of motivations (and constraints) that women identify.

The implications of this study for theorising slow travel and tourism are many. First, we need to move beyond the notion that slow travel experiences are premised on an implicitly masculine subject and examine how gender plays out through different experiences of mobility. Second, the understanding of social and identity aspects of the slow experience needs further exploration alongside the environmental concerns (as well as the relationship between them). These issues call for a greater engagement in diverse methodologies through which experiences of slow mobility can be better understood (Büscher & Urry, 2009). Within the context of the slow movement, the practices of travel and tourism are a growing dimension of our contemporary 'life politics' through which alternative cultural and political forms of hedonism are played out. Slow mobilities may well move from being considered niche or alternative to becoming mainstream as the world grapples with issues of peak oil, increasing international demand for travel and problems of food security (Sheller & Urry, 2006).

References

Alvesson, M. and Skoldberg, K. (2000) *Reflexive Methodology: New Vistas for Qualitative Research*. London: Sage.

Australian Bureau of Statistics (2010) *Participation in Sport and Physical Recreation, Australia, 2009–10* (cat. no. 4177.0). Canberra: Australian Bureau of Statistics.

Bicycle Queensland (2009) *Bicycle Queensland Annual Report 2009*. http://www.bq.org.au/_downloads/BQAnnualReport2009.pdf:Brisbane

Bonham, J. and Koth, B. (2010) Universities and the cycling culture. *Transportation Research Part D: Transport and Environment* 15 (2), 94–102.

Büscher, M. and Urry, J. (2009) Mobile methods and the empirical. *European Journal of Social Theory* 12 (1), 99–116.

Cresswell, T. (2010) Towards a politics of mobility. *Environment and Planning D: Society and Space* 28 (1), 17–31.

Dickinson, J. and Lumsdon, D. (2010) *Slow Travel and Tourism*. London: Earthscan.

Dickinson, J., Lumsdon, L. and Robbins, D. (2010) Slow travel: Issues for tourism and climate change. *Journal of Sustainable Tourism* 19 (3), 281–300.

Dickinson, J. and Robbins, D. (2009) 'Other people, other times and special places': A social representations perspective of cycling in a tourism destination. *Tourism and Hospitality Planning and Development* 6 (1), 69–85.

Emond, C., Tang, W. and Handy, S. (2009) Explaining gender difference in bicycling behavior. *Transportation Research Record: Journal of the Transportation Research Board* 2125(1), 16–25.

Furness, Z. (2010) *One Less Car: Bicycling and the Politics of Automobility*. Philadelphia, PA: Temple University.

Garrard, J. (2003) Healthy revolutions: Promoting cycling among women. *Health Promotion Journal of Australia* 14 (3), 213–215.

Garrard, J., Rose, G. and Lo, S. (2008) Promoting transportation cycling for women: The role of bicycle infrastructure. *Preventive Medicine* 46 (1), 55–59.

Germann Molz, J. (2009) Representing pace in tourism mobilities: Staycations, Slow Travel and *The Amazing Race*. *Journal of Tourism and Cultural Change* 7 (4), 270–286.

Germov, J., Williams, L. and Freij, M. (2011) Portrayal of the Slow Food movement in the Australian print media. *Journal of Sociology* 47 (1), 89–106.

Hall, C.M. (2009) Degrowing tourism: Décroissance, sustainable consumption and steady-state tourism. *Anatolia: An International Journal of Tourism and Hospitality Research* 20 (1), 46–61.

Hanson, S. (2010) Gender and mobility: New approaches for informing sustainability. *Gender, Place & Culture: A Journal of Feminist Geography* 17 (1), 5–23.

Hodgson, L. (2007) Bicycle touring and the construction of place: A study of place meanings on a cycle tour of Tuscany, Italy. In F. Jordon, L. Kilgour and N. Morgan (eds) *Academic Renewal: Innovation in Leisure and Tourism Themes and Methods* (Vol. 97, pp. 25–41). Eastbourne: Leisure Studies Association.

Holloway, I., Brown, L. and Shipway, R. (2010) Meaning not measurement: Using ethnography to bring a deeper understanding to the participant experience of festivals and events. *International Journal of Event and Festival Management* 1 (1), 74–85.

Jackman, C. (2009) 7 days to save the planet. *The Weekend Australian Magazine*, 19 December, p. 20.

Lamont, M. and Causley, K. (2010) Guiding the way: Exploring cycle tourists' needs and preferences for cycling route maps and signage. *Annals of Leisure Research* 13 (3), 497–522.

Little, D. and Wilson, E. (2005) Adventure and the gender gap: Acknowledging diversity of experience. *Society and Leisure* 28 (1), 185–208.

O'Connor, J. and Brown, T. (2007) Real cyclists don't race: Informal affiliations of the Weekend Warrior. *International Review for the Sociology of Sport* 42 (1), 83–97.

Parkins, W. and Craig, G. (2006) *Slow Living.* Sydney: UNSW Press.

Pritchard, A. and Morgan, N.J. (2000) Constructing tourism landscapes – Gender, sexuality and space. *Tourism Geographies: An International Journal of Tourism Space, Place and Environment* 2 (2), 115–139.

Ritchie, B., Tkaczynski, A. and Faulks, P. (2010) Understanding the motivation and travel behavior of cycle tourists using involvement profiles. *Journal of Travel and Tourism Marketing* 27 (4), 409–425.

Rojek, C. (2010) *The Labour of Leisure.* London: Sage.

Rose, N. (1999) *The Powers of Freedom: Reframing Political Thought.* Cambridge: Cambridge University Press.

Sheller, M. and Urry, J. (2006) The new mobilities paradigm. *Environment and Planning A* 38, 207–226.

Simpson, C. (2001) Respectable identities: New Zealand nineteenth-century 'new women' – on bicycles! *The International Journal of the History of Sport* 18 (2), 54–77.

Soper, K. (2008) Alternative hedonism, cultural theory and the role of aesthetic revisioning. *Cultural Studies* 22 (5), 567–587.

Spinney, J. (2009) Cycling the city: Movement, meaning and method. *Geography Compass* 3 (2), 817–835.

Tam, D. (2008) Slow journeys: What does it mean to go slow? *Food, Culture and Society: An International Journal of Multidisciplinary Research* 11 (2), 207–218.

Uteng, P.T. and Cresswell, T. (eds) (2008) *Gendered Mobilities.* Aldershot: Ashgate.

Watts, L. and Urry, J. (2008) Moving methods, travelling times. *Environment and Planning D: Society and Space* 26, 860–874.

Womack, C. and Suyemoto, P. (2010) Riding like a girl. In J. Ilundain and M. Austin (eds) *Cycling Philosophy* (pp. 81–93). Chichester: Wiley Blackwell.

9 Wandering Australia: Independent Travellers and Slow Journeys Through Time and Space

Marg Tiyce and Erica Wilson

> We're heading around the country...but there's no rush. We're just wandering along at our own pace, taking our time and smelling the roses. Every minute is precious to us and we're determined to make the most of it. (Sylvie, study participant)

Wandering, drifting, exploring, walking, ambling, meandering. As Sylvie's opening comments suggest, these forms of travelling readily convey a sense of slowness in time and space and evoke notions of mobilities that are purposefully and mindfully unhurried and self directed. This chapter explores the concept of slow travel through an ethnographic study of the experiences of a group of long-term independent travellers in Australia. Describing themselves as 'wanderers', these travellers enact journeys that are distinct from other touristic mobilities in that they consciously travel slowly, engaging with time and space in very different ways. In exploring these travellers' slow wanderings, we draw on contemporary conceptualisations of slow living and the slow movement, which are 'based on a questioning of the conventional wisdoms regarding the temporal and spatial contexts of global culture' (Parkins & Craig, 2009: 9).

Time and space are powerful bases of social practice in contemporary Western society, structuring, strengthening and securing the conduct of our social and personal lives (May & Thrift, 2001; Urry, 1996). Yet, these modern conceptualisations of time and space have been widely deconstructed as artificial, culturally determined and human made, devised to satisfy industrial imperatives of productivity, efficiency and control (Bluedorn, 2002; Harvey, 1989; Lauer, 1981; Levine, 1997). Rose (1999) notes that capitalist societies and neo-liberal governments require practices of the self to be organised around human-made devices and 'assemblages' of technologies, so that individual conduct can be shaped and controlled, and certain social outcomes achieved. In their simplest forms such devices include timepieces, calendars, media, information technologies and the surveillance of their use (Levine, 1997; May & Thrift, 2001; Rose, 1999). These structures and technologies are used to regulate and control the conduct of most aspects of our everyday lives (Massey, 2005; May & Thrift, 2001; Urry, 1996). They have also become an integral part of our lives, enhancing our relationships and connectedness with each other. Nevertheless, these ever developing technologies, Rojek (1999) argues, have contributed to a perceived 'speed' and pace of modern life that can be both challenging and all consuming.

Discussions about the fast pace of modern life, time poverty and the endless search for work–life balance permeate modern lives (Langley & Breese, 2005; Shaw, 2001). According to Virilio (1977), the ever increasing accessibility and proximity that technologies afford results in dissolution of boundaries and creates a sense of time-space 'compression'. Under such conditions, demands on the individual can challenge their capacity to function and cope (Rose, 1996). Nevertheless, the productive use of time and space is considered a vital responsibility for each individual, such that, along with the demands of productivity and performance, even personal and leisure time are now mediated by strong cultural expectations and controls, and made into work (Rojek, 1999; Rose, 1996; Veal, 1987, 2009). Outside the home, few spaces are free of these social expectations and their public scrutiny (Altman & Werner, 1985; Mallett, 2004) and even home is recognised as an increasingly demanding and busy space for work (Dovey, 1985; Dupuis & Thorns, 1998; Giddens, 1991). As Parkins and Craig (2009: 1) lament, we are now suffering from the uniquely 'late modern malaise' of 'hurry sickness'.

Parkins and Craig (2009: 1) argue that just 'as speed is seemingly equated with efficiency and professionalism … slowness can become a way of signalling an alternative set of values or a refusal to privilege the workplace over other domains of life'. The slow movement is a direct response to dominant, capitalist constructions of time, space, production and consumption (Parkins & Craig, 2009). Practices of time and space are political as well as cultural and historical constructs (Massey, 2005). As evolving technologies gradually erode time-space boundaries, understandings of space and distance are also increasingly challenged (Massey, 2005; May & Thrift, 2001; Virilio, 1977). Ideas of the world as an ever closer and more accessible 'global village' open up alternative spatial and temporal opportunities (Harvey, 1989: 240), especially for those who travel.

Travelling in order to slow down the pace of life is certainly not new. Desires for slower and quieter spaces and places for quiet relaxation and recreation underlie much travel (Krippendorf, 1987). Travelling independently allows greater self-determination, flexibility and personal control over the journey (Elsrud, 1998). Coming and going when one likes gives an individual personal control over their time and space, and hence a greater sense of control over their lives (Levine, 1997; Mehmetoglu *et al.*, 2001). Similarly, travelling long term obtains greater release from social demands, obligations and scrutiny. Such travel is centred on enjoying 'the good life', a slower, easier and more personally directed way of life (Shaw, 2001). Travel, after all, is intrinsically spatial and temporal in nature (Massey, 2005). As a form of slow living, slow travel embraces practices of time and space that emphasise awareness, mindfulness, pleasure and care (Parkins & Craig, 2009). Little research, however, has examined these constructions of time and space in the context of contemporary, long-term independent travellers, particularly wanderers on their journeys around Australia.

Ethnography of Wandering Travellers

The following discussion draws on the findings of ethnographic research conducted by the first author on the experiences of independent long-term travellers in Australia. Concerned with digging beneath the layers of accepted tourism discourses, the study sought to gain in-depth phenomenological understandings of travellers' experiences from their perspective (Hammersley & Atkinson, 1992; Weber, 1978). A contemporary postmodern lens emphasised the uniqueness, diversity and dynamism of individual experiences and meanings within the broader social context of their lives (Gubrium & Holstein, 1997).

Ethnographic participation in the travelling culture and way of life over a five-year period enabled the researcher a subjective positionality, allowing her to become enmeshed in the experiences and everyday lives of fellow travellers (Denzin & Lincoln, 2005; Geertz, 1973). Travelling alongside these wanderers (sharing communal spaces, resources and experiences) allowed meaningful exchanges to arise in a 'naturalistic' way (Lincoln & Guba, 1985). Sixty-two in-depth interviews and over 200 semi-structured conversations were recorded with solo, partnered and group travellers aged from 18 to 82 years. A constructivist, grounded approach to analysis highlighted recurring themes and processes, and these were used to form interpretations and propositions, while privileging participants' views, understandings and voices (Charmaz, 2006; Glaser & Strauss, 1967; Hammersley & Atkinson, 1992; Reinharz, 1992). This chapter discusses one of these key themes, the desire for slow travel and some of the ways travellers shape time and space to achieve it.

Wandering Through Time and Space

Wandering as a form of travelling has a long history that reveals a preference for a different, slower way of moving through the world. Cohen (1973) notes that contemporary drifters are experimenting with alternative lifeways in a quest for greater meaning in life. Travellers described in this study emphasised that wandering represents an alternative and unconventional form of travelling, signalling their desire to move away from more popularised forms of organised travel and tourism. As such, wandering represents a form of touring without fixed itineraries, its key experiences made accessible by consciously travelling slowly and in a personally determined way. As two participants suggested:

> When I say I'm a wanderer, I mean I'm not in a hurry to get anywhere. I'm not interested in all that rushing about. ... All those tourist places don't interest me. I'm just quietly poking about mate, doing what interests me. It doesn't matter where I end up. (Max)

> This is not about going somewhere in particular, it's definitely not like a holiday. It's about having a life, changing the way we do things day to day, having a better kind of life. It's the good life we're after if we can find it. (Tom)

Yet wandering is more than travelling slowly and in a more fulfilling way. For these travellers, wandering is about claiming time and space, about moulding and folding time and space in their own way in order to enhance their experiences and their sense of well-being. This wandering represents a different way of living and being in the world, inciting notions of resistance to established performances of tourism bound on achieving destinations and consuming holidays.

> We haven't got loads of cash but even if we did we would still be just hanging out camping and fishing and chatting to people. The simple things are the most rewarding ... and having the time to relax and enjoy them. (Joy)

While independent travellers are often categorised as 'budget' travellers, these wanderers are not especially driven by budgets. Rather, they shun consumerist beliefs and patterns of touristic behaviour that equate spending money with having quality leisure and life experiences.

> Arh, the good life. ... No amount of money can buy a view like that [another glorious desert sunset]; wide open spaces, a few good friends and the horizon stretching out as far as the eye can see. ... Whose idea was it that you have to work your life away always chasing the next best, latest and greatest gadget? (John)

For travellers like John, the best experiences are simple ones that align them more closely with self and nature, such as 'driving through barren and vast landscapes', or 'quietly viewing the sun setting over the desert horizon' or 'sharing the warmth and camaraderie of fellow travellers around a campfire'. Living simply and closely with nature is highly valued by these travellers. There is indeed an inherent romanticism about these idealised natural experiences, but also an awakening consciousness around self and the world. They embrace the contrasts between city and country, coast and desert, settlement and bush, ruminating over human footprints and questions of environmental integrity, human habitation and negative human impact. Issues of congestion and degradation are juxtaposed with experiences of space and 'naturalness' and call into question cultural, social and personal ideals.

Wanderers adopt a wide range of travel means including carrying a tent or swag, towing a caravan, or driving a campervan or motorhome. No matter how modest or luxurious the vehicle, living on the road, of necessity, involves a commitment to the experience of living with less and desisting from the accumulation of material possessions (at least while travelling). Such an approach favours living more simply, unencumbered by consumerist undertakings aimed at purchase and possession of ever increasing objects and reputation in the name of home, what Goffman (1967) suggested were the 'props' of success. In this way, travelling independently and long term signals both a restlessness and pragmatism around home and the conduct of everyday life. Rather than having a desire to possess, they move to slough off that which would hold them in a sense of insufficiency and striving for more.

For these travellers, one home is left behind, but a new home is simultaneously created 'on the road'. Home in this sense is peeled back to its simplest bounded form – shelter, store, refuge; a space of safety, sustenance and belonging (Dovey, 1985; Mallett, 2004; Saunders & Williams, 1988). Travellers evade attachment to places, since home as residence locates and fixes people in time and space (Mallett, 2004). As bell hooks (1991: 148) notes 'home is no longer one place. It is locations.' For wanderers home is located in many alternative ways (for example, home is country, home is family, home is within). Sylvie showed her ambivalence to home by questioning:

> What is a home after all? Home is where the wind takes me. As long as I have health and happiness, what do I want with a house? This life is my home now (Sylvie).

In these ways, travelling long term and living on the road can be viewed as culturally and politically radical, a space for claiming sovereignty and contesting collective ideals and hegemonies that normalise everyday lives around notions of house as home. Wandering clearly disrupts such structures. The supposed distinction between home and away has also been questioned by cultural geographers (Curtis & Pajaczkowska, 1994). Indeed, a common motorhome insignia reads 'No Fixed Address', challenging socially idealised and reinforced standards of home as house, family, community and settlement. Rather than conforming to social norms of home as fixed in time and space, travellers actively embrace mobility and home on the road in order to shape their life on their own terms. As one veteran traveller asserted, wandering is 'a state of mind, an attitude, and a darn determination to do things your own way' (Fred).

Pace: Taking our Own Sweet Time

Long-term wanderings allow travellers greater control over the pace of their journeys. Wandering means travellers are able to move at their own pace, guided by what nurtures, and is rewarding and personally meaningful. The pace of journeys is a significant issue, one which they monitor and seek to control in order to sustain their energy and to ensure that everyday life contributes to their sense of well-being. As travellers shared:

> Look, we're always in a hurry to see what's over the next horizon, that's human nature. So we watch our mileage, aim for double digit days [below one hundred kilometres] even when we're out in the desert we try to plod along, slow and easy. Otherwise you miss out on too much. Going slow means you've got plenty of time for living.... It's living the good life. (Geoff)

> It's a joy to be wandering along, a real joy, stopping whenever we see something nice, taking our own sweet time. There's time to listen to the universe, to hear its stories and messages reaching out to us. Staying over when people are nice and hearing their stories and giving back to them. You can dawdle over things, linger longer as they say.... It brings a feeling of abundance into your life. There's more harmony and balance, you become more centred and whole, I think. (Mary)

The slower pace of wandering provides opportunities for living the good life; for dawdling, hearing, listening and seeing. As Virilio (1977) emphasises, the speed of experiences and happenings changes their essential nature. Slowing down, for these travellers, creates time and space to experience and appreciate the journey – to engage with and enjoy other people and places, to immerse themselves in nature and self, to pursue special interests, to share their experiences with like-minded travellers. These experiences of slow living involve using time more attentively, more mindfully, and with concern for pleasure and purpose (Parkins & Craig, 2009). In this way, wandering travels go beyond the notion of a bounded holiday to contribute to travellers' wider lives in ways that are personally fulfilling and meaningful – providing 'memories', 'the good life', 'a feeling of abundance', 'harmony and balance'.

It's the in-between places that you enjoy the most when you are travelling, the things you stumble across by chance, and can stop and explore. Some of my best memories are from along the side of the road, an old tree or getting close up to a wild emu or stopping for a cuppa and a chat with other travellers. (Peter)

Travellers, like Peter, reject the prime importance of places in travel, preferring the 'in-between' and having the time to simply 'stop and explore'. For these travellers, places are 'integrations of time and space; *spatio-temporal events*' (Massey, 2005: 130, emphasis in original). The idea that space is static (May & Thrift, 2001) is rejected, travellers move readily between places viewing space, like time, as fluid and dynamic, little more than a signifier of progress.

Broome was just the turn around point for us.... The rest was one big continuous adventure. (Mima)

Wandering along out here [in the desert] you lose all track of time. Moving along through millions of years of changing landscapes, like rolling along in a slipstream, living its changing, evolving moods.... The surrealness of a world without things to hold you down in one place. (Michael)

For wanderers, 'space and time are inextricably interwoven' (Massey, 1994: 261). Wandering is a stream of slowly unfolding events and experiences often not framed and defined by human constructions. In this fluid 'timespace' (May & Thrift, 2001), human made constructions of time and space (especially those that would bind and obstruct) are largely abandoned and more natural markers adopted. The rhythms of nature, seasons, day and night, provide markers and replace demarcations of watches and calendars. People move at their own pace, guided by what feels personally easy and comfortable, rewarding and nurturing.

It's so freeing to have the time and space to just be me and do what I need for myself. That's not something I think I've ever really had before. There's always been family to care for and things to think of. It's always been 24/7. Just for once, not having to be responsible to the rest of the world, not having to answer to anyone but myself. (Julie)

Julie's resistance to being caught up in the dramas and dilemmas of other people's lives is evident. In many ways, Julie is resisting conventional notions of how time and space are often defined for women through the 'ethic of care' – caring, thinking about and helping others (Deem, 1996; Wilson & Little, 2005). Wearing and Wearing (1988) note women's experiences of leisure are often mediated by domestic obligations and subject to strategies of outdoor exclusion (Rojek, 1999). As a solo woman traveller, Julie's travel was less constrained than many partnered women travellers, for whom expectations of domesticity remained and were often a 'sticking point', as Lisa explained:

> It's a bit of a sticking point, but I find my way around it. He still expects me to do the cooking and cleaning, even in a tent ... and I'm the one that rings the kids. But I'm working on it ... like sometimes I jack up and tell him it's his turn. (Lisa)

For many women in this study, travelling offered opportunities to resist and rewrite established gender roles around 'work'. The slower pace of their lives and the greater time and space it afforded assisted such endeavours. Often though, it was only in taking time away from domestic and family responsibilities that they shaped time and space for themselves (Deem, 1996; Elsrud, 1998). Resistance to social expectations was not only limited to women though. For example, Wayne talked about how he had:

> ... pretty much given my life to looking after the family, the business, and anyone else who came along needing help. There's so much I want to do, too, but never have the time.... [Travelling means] finding some space and time for all that. I've earned it I think. (Wayne)

Making time and space through travel creates social distance and untangled, unencumbered spaces away from the demands, distractions and constraints of more normal and complex social relationships and lives. For women, a sense entitlement allows them to begin to prioritise themselves (Henderson & Bialeschki, 1991). For men too, attitudes of entitlement were important in enabling them to throw off their previously obligated lives, and prioritise themselves and their personal needs and desires.

For these travellers, this involves taking and expanding personal time and space for slow and leisurely enjoyment of their day-to-day lives. They are happy to achieve less – to idle, to relax and to enjoy a simpler daily life. In these ways, travellers argued, life evolves 'more naturally' and is made more meaningful and enjoyable in the process. Joe suggested:

> This is how nature intended us to live, I reckon. Not forcing things to happen but just letting life happen more naturally. When you watch the birds or [other animals] you see a different side of life, a more natural agreeable way of going through life, ... too much is expected of people these days. (Joe)

For Joe, his previous life 'of overwork and no play' was an unnatural way of moving through life. It was taxing on his sense of health and well-being, and he believed 'it was asking too much'. Acutely aware of the demanding character of their previous lives, these travellers firmly resisted speed and planning as unsatisfying ways of living, preferring as Joe insisted, 'the slow lane'.

To this end, travellers tend to allow their journeys and experiences to unfold serendipitously. Through this gentle unfolding of daily life, they embrace uncertainty in return for elements of surprise and adventure that promise enhanced wonder, novelty and enrichment of life. They favour moving slowly through landscapes and communities (rather than dashing across or through them); taking time to stop, share, observe and 'chew the fat' (chat) with locals or other like-minded travellers. They seek deeper appreciation and understanding of life through their slow and accidental encounters and meaningful connections with people and places, their stories and imaginings – through their 'meeting up of histories' (Massey, 2005). They wonder how their own dreams and stories might belong amongst these other ways of knowing and doing and living. Fullagar (2001) observes that this openness and wonder can evoke a sense of life's abundance and provide a lens into other ways of knowing and appreciating.

For all travellers, the simple unfolding of serendipitous experiences was found to be more fulfilling and meaningful than the accepted prudence of organised and scheduled touristic holidays.

> We dashed across the Nullarbor on our first trip. We thought, from what people had told us that it was the sort of place you got through as quick as you could, a no-mans land. This time we took it slow and meandered across. And what a beautiful place it turned out to be. So alive with wildlife and fantastic sunsets. And the coastline there ... it was such a special part of our trip. (Jeanie)

Slowing down, it seems, means time to look and see, to appreciate and wonder. For Jeanie, and many other travellers, taking time could transform experiences of space – a 'no-man's land' was transformed into a space 'alive' and 'fantastic' and 'beautiful'. As Joe explained 'when you're wandering you have the time to look deeply into the crevices', to explore the 'secret and mysterious' spaces. Such wonder, Fullagar (2001: 301–302) believes, 'has the potential to move us beyond the fixed boundaries of self as separate from the world, and into an ethical relation of openness with otherness both within and beyond the self'. In this way, notions of serendipity are bound up with discovery and wonder, but also with a greater realisation of self and self in the world.

In the Moment: Privileging the Here and Now

Slow travel is also about living actively and fully in the present. Travellers' aversion to planning denotes a conscious strategy to 'leave tomorrow to its own devices', as one traveller, Dean, claimed. Tomorrows became distant contrivances and uncertainties. Tomorrows, like the lives left behind, were seen as imbued with needless workings and strivings. Instead, these travellers longed to be free of the issues of yesterday and tomorrow and found a sense of freedom in abandoning the strictures of time and space, and in actively privileging the 'here and now':

> It's here and now that's important to me now. I can't be bothered with what happened yesterday or who said what to who, or what might happen tomorrow for that matter. I've had enough of all that crap. Today is what counts. (Leanne)

Partitioning of past and future times and spaces provided a sense of release and escape, yet it was the freedom that accompanied these timespace practices that travellers truly relished. Aaron's comments were typical:

> There's a lot of freedom in living day to day.... One day simply follows another and you don't need to look ahead and worry about where you gotta be and what you gotta do.... You can truly make each day count. (Aaron)

One of the most notable features of the way these wanderers experience their lives is their orientation towards 'living in the moment' or living in the 'now' – the present. The longer people travel, it was suggested, the more they tend to abandon thinking about the past and future in favour, as Gayle said, of a 'life as it unfolds':

> I am usually a fanatical planner but since I've been out here I just roll along and live life as it unfolds. Living in the moment. For me it's the great joy of travelling this way, it's such a wonderfully exciting but also calming way of life. All the stress and worry is put behind you and you leave yesterdays and tomorrows alone and just go with the flow. That's a growing experience I think, for me a better way to deal with what life hands out. There's a greater feeling of contentment and wellness. (Gayle)

For Gayle and many other travellers, feelings of 'contentment and wellness' came from their practices of living in the moment. In this timespace, past and future fade from view in favour of living in the moment and appreciating each experience as it unfolds. Like Csikszentmihalyi's (1975) conceptualisation of flow, experiences of living in the moment are described as altering perceptions of time and space such that people become detached from the past and the future, and from external happenings. They are then intensely immersed in, and engaged with, their present experiences no matter how small or insignificant those experiences might seem. Slowness provides space for such intensity of engagement and experience.

Practices of living in the moment and travelling slowly challenge neoliberal ideals that assume individual striving is the means to 'becoming' fulfilled, as Giddens' (1991) modern reflexive project of the self would suggest. Rather than continually striving to improve themselves and their social positions in the world, these travellers' orientation towards living in the moment related more closely with preferences for simply 'being' in the world (Heidegger, 1980). Travelling slowly provided opportunities for periods of intense calm, for release and letting go. Contentment was found in the freedoms and ordinariness of appreciating what is, of simply being themselves and enjoying each moment and each situation as it unfolds. In this way, experiences of simply being could be deeply transformational in that they demonstrated simple truths about life and self that decentred bounded identities and gave space for redefined notions of identity and self.

It opened my eyes to what's important to me and to the realisation that I am okay as I am; that I don't have to be always working at something, building things and improving myself. I am okay, I belong here. I can just be me and that's enough. (Mary)

We were so far away from the world today, just the water and us. Nothing can touch us. (Jeanie)

Their physical detachment and distance from the centre of society *and* its cultural mores, what Simmel (1993) would call their *outsiderness*, certainly facilitates their shift to experiences and practices of living in the moment and going slow. Yet, so too does the contentment and stillness of the mind that results from experiences of simply being, of not thinking, not planning, and not living in the demanding and constraining spaces of yesterdays and tomorrows. As Julie said:

It's good to shake off the memories and the worries. Your mind starts to slow down and you become calmer and happier on the inside. I've got this stillness of mind now that I never knew was possible. There are long periods of time in each day now when I'm engrossed in the world around me and I'm not thinking at all. Beachcombing and swimming are my new passions, there's no real mental work to them, no pushing to do something, I'm just there idling along as happy as can be. (Julie)

To 'shake off' these more egoic thinking processes (as past and further imaginings always are) in favour of present unconscious feelings and sensory experiences is to open up time and space for more self orientated experiences, especially of pleasure and play (Krippendorf, 1987; Tolle, 1999). Jung avowed these timespaces of 'creative play' encouraged healing and harmony, and recovery and re-creation of body, mind and spirit (Jung, 1995).

Spending Precious Time

The preciousness of time and the need to make every minute count features strongly in the stories and conversations of these long-term wanderers. And while the notion of wandering might suggest indulgences in idleness and the casual wasting of time, travellers emphasise that, in their view, little time is wasted on the road. Instead, time is recognised as a highly precious and scarce resource to be spent on experiences that enhance their everyday lives. As Ron suggested:

> You're not just wasting your life away on work and running around. Every minute counts out here, even when you're sitting in a chair and gazing out to sea. It's important to have that time. The thing is, it's precious time because you can spend it however you want to. (Ron)

> Having time to come out here and do some fishing and lazing around in the great outdoors, well it's good for the body and the mind. A way of getting over all that ails you, I reckon.... You're too long dead after all. (Wayne)

For many travellers, time was acutely more precious when viewed in relation to their own mortality. 'The clock is ticking', Joe declared 'and who knows when it will stop'. Uneasy conversations emerged around the passage of time and the limited time each person may have left in their life. Such uncertainties lead to a desire to use 'as much time as possible' in meaningful and self-directed ways. Gloria suggested 'it brings to light all that's important ... [and] highlights the need to live life to the full while you still can'. Wanderers have a strong belief in the practices of travelling slowly to enhance their well-being and the quality of their lives (Tiyce, 2008).

Issues of mortality provoked a common concern, not just with spending precious time wisely and well, but with extending the quality, depth and meaningfulness of their lives through ideas and practices of living well. Wanderers view moving slowly through the world as critical to living well and living the good life. Their slow wanderings and altered timespace practices represent not only their resistance to conventional ill-fitting ways of living but also their search for a better, slower way of living and being in the world.

Conclusions and Final Wanderings

This discussion has explored the way a group of long-term independent Australian travellers use time and space to shape slow journeys and slower, more meaningful lives on the road. Travelling slowly, these wanderers consciously shape their experiences and timespace practices in ways that nurture and enhance their sense of well-being and the quality of their lives. They allow their journeys to unfold serendipitously, revelling in the detail and the surprise. They privilege the present, over the past and future, and more often follow timespace markers of nature, events and experiences. These attitudes and practices allow for a slower pace of life and a life more meaningful, what they perceive as 'the good life'.

As a form of slow living, this slow travel privileges self and challenges socio-political hegemonies that regulate and control the use of personal time and space (Parkins & Craig, 2009). Such hegemonies insist on subordination of personal needs and desires in the interests of productivity, participation, social cohesion and other goals. However, wanderers refute modern ways of living that view busyness, performance, productivity and pace as measures of success and fulfilment. They favour a simple life on the road, free of the demands of a life constrained by the regulation of time and space in conventional society. As these travellers suggest, wandering, like other slow travel movements, defies normative ways of living and being in contemporary societies and expresses a yearning for, and an expression of, alternative, slower and richer ways of living and being in the world.

Urry (1999) conceives of a world moving in a realm of continuous and ever-increasing mobility. If these travellers and their slow wanderings are an example of this shift, such a world has the potential to change the very nature and conduct of social life itself, not just the nature and experience of time and space through practices of slow travel.

References

Altman, I. and Werner, C. (1985) *Home Environments.* New York: Plenum Press.

Bluedorn, A. (2002) *The Human Organisation of Time: Temporal Realities and Experience.* Stanford, CA: Stanford Business Books.

Charmaz, K. (2006) *Constructing Grounded Theory: A Practical Guide Through Qualitative Research.* Thousand Oaks, CA: Sage.

Cohen, E. (1973) Nomads from affluence: Notes on the phenomenon of drifter-tourism. *International Journal of Comparative Sociology* 14 (1–2), 89–103.

Csikszentmihalyi, M. (1975) *Beyond Boredom and Anxiety.* San Francisco, CA: Jossey-Bass.

Curtis, B. and Pajaczkowska, P. (1994) 'Getting there': Travel, time and narrative. In G. Robertson, M. Mash, L. Tickner, J. Bird, B. Curtis and T. Putnam (eds) *Travellers' Tales: Narratives of Home and Displacement* (pp. 199–215). London & New York: Routledge.

Deem, R. (1996) No time for a rest? An exploration of women's work, engendered leisure and holidays. *Time & Society* 5 (1), 5–25.

Denzin, N.K. and Lincoln, Y.S. (2005) *Handbook of Qualitative Research* (3rd edn). Thousand Oaks, CA: Sage.

Dovey, K. (1985) Homes and homelessness. In I. Altman and C. Werner (eds) *Home Environments* (pp. 33–64). New York: Plenum Press.

Dupuis, A. and Thorns, D.C. (1998) Home, home ownership and the search for ontological security. *Sociological Review* 46 (1), 24–47.

Elsrud, T. (1998) Time creation in travelling: The taking and making of time among women backpackers. *Time & Society* 7 (2), 309–334.

Fullagar, S. (2001) Desire, death and wonder: Reading Simone de Beauvoir's narratives of travel. *Cultural Values* 5 (3), 289–305.

Geertz, C. (1973) *The Interpretation of Cultures: Selected Essays.* New York: Basic Books.

Giddens, A. (1991) *Modernity and Self Identity: Self and Society in the Late Modern Age.* Cambridge: Polity Press.

Glaser, B.G. and Strauss, A.L. (1967) *The Discovery of Grounded Theory: Strategies for Qualitative Research.* Chicago, IL: Aldine.

Goffman, E. (1967) Where the action is. In *Interaction Ritual: Essays on Face to Face Behaviour.* New York: Doubleday.

Gubrium, J.F. and Holstein, J.A. (1997) *The New Language of Qualitative Method.* Oxford: Oxford University.

Hammersley, M. and Atkinson, P. (1992) *Ethnography: Principles in Practice.* London: Routledge.

Harvey, D. (1989) *The Condition of Postmodernity: An Inquiry into the Origins of Cultural Change.* Oxford: Blackwell.

Heidegger, M. (1980) *Being and Time* (J. Macquarie and E. Robinson, trans.). Oxford: Blackwell (original work published 1927).

Henderson, K. and Bialeschki, M.D. (1991) A sense of entitlement to leisure as constraint and empowerment for women. *Leisure Sciences* 13 (1), 51–65.

Hooks, B. (1991) *Yearning, Race, Gender and Cultural Politics.* London: Turnaround.

Jung, C. (1995) *Memories, Dreams, Reflections.* Recorded by A. Jaffe (ed.) (R. Winston and C. Winston, trans.). London: HarperCollins (original work published 1963).

Krippendorf, J. (1987) *The Holiday Makers: Understanding the Impact of Leisure and Travel.* Oxford: Heinemann Professional.

Langley, C. and Breese, J. (2005) Interacting sojourners: A study of students studying abroad. *The Social Science Journal* 42 (2), 313–321.

Lauer, R. (1981) *Temporal Man: The Meaning and Uses of Social Time.* New York: Praeger.

Levine, R. (1997) *A Geography of Time*: On Tempo, Culture and The Pace of Life. New York: Basic Books.

Lincoln, Y.S. and Guba, E.G. (1985) *Naturalistic Inquiry.* Newbury Park, CA: Sage.

Mallett, S. (2004) Understanding home: A critical review of the literature. *The Sociological Review* 52 (1), 62–84.

Massey, D. (1994) *Space, Place and Gender.* Cambridge: Polity Press.

Massey, D. (2005) *For Space.* London: Sage.

May, J. and Thrift, N. (2001) *Timespace: Geographies of Temporality.* London: Routledge.

Mehmetoglu, M., Dann, G. and Larsen, S. (2001) Solitary travellers in the Norwegian Lofoten Islands: Why do people travel on their own? *Scandinavian Journal of Hospitality & Tourism* 1 (1), 19–37.

Parkins, W. and Craig, G. (2009) *Slow Living.* Oxford: Berg.

Reinharz, S. (1992) *Feminist Methods in Social Research.* Oxford: Oxford University.

Rojek, C. (1999) *Decentring Leisure: Rethinking Leisure Theory.* London: Sage.

Rose, N. (1996) *Inventing Ourselves: Psychology, Power and Personhood.* Cambridge: Cambridge University Press.

Rose, N. (1999) *Powers of Freedom: Reframing Political Thought.* Cambridge: Cambridge University Press.

Saunders, P. and Williams, P. (1988) The constitution of the home: Towards a research agenda. *Housing Studies* 3 (2), 81–93.

Shaw, J. (2001) Winning territory: Changing place to change pace. In J. May and N. Thrift (eds) *Timespace: Geographies of Temporality* (pp. 120–132). London: Routledge.

Simmel, G. (1993) The stranger. In C. Lemert (ed.) *Social Theory: The Multicultural and Classic Readings* (pp. 200–204). Boulder, CO: Westview.

Tiyce, M. (2008) Healing through travel: Two women's experiences of loss and adaptation. In S. Richardson, L. Fredline, A. Patiar and M. Ternel (eds) *Proceedings of the 18th Annual CAUTHE Conference.* Gold Coast, 11–14 February.

Tolle, E. (1999) *The Power of Now.* Sydney: Hodder.

Urry, J. (1996) Sociology of time and space. In B.S. Turner (ed.) *The Blackwell Companion to Social Theory* (pp. 369–395). Oxford: Blackwell Publishers.

Urry, J. (1999) *Sociology beyond Societies: Mobility for the Twenty-First Century.* London: Routledge.

Veal, A.J. (1987) *Leisure and the Future.* London: Allen and Unwin.

Veal, A.J. (2009) The elusive leisure society (4th edn) (School of Leisure, Sport and Tourism Working Paper 9). Sydney: University of Technology Sydney – Online document: http://datasearch.uts.edu.au/business/publications/1st/index.cfm

Virilio, P. (1977) *Speed and Politics: An Essay on Dromology.* New York: Semiotexte.

Wearing, B. and Wearing, S. (1988) All in a day's leisure: Gender and the concept of leisure. *Leisure Studies* 7 (2), 111–123.

Weber, M. (1978) *Economy and Society: An Outline of Interpretive Sociology* (G. Roth and C. Wittich, eds). Berkeley, CA: University of California Press.

Wilson, E. and Little, D.E. (2005) A 'relative escape'? The impact of constraints on women who travel solo. *Tourism Review International* 9 (2), 155–175.

10 Alternative Mobility Cultures and the Resurgence of Hitch-hiking

Michael O'Regan

While tourist mobilities occur at a range of spatial scales, the ways in which tourists move, dwell and communicate have been understudied and undertheorised. Their geographic movement has often been stereotyped as being produced by an embodied tourist habitus that is continually developing within a hegemonic tourist culture. Tourist mobilities are related 'practices' that are strategically engineered through powerful media discourses, often intensified by transport and service providers who offer access to peoples, places and cultures in all corners of the globe. This chapter, however, argues that individuals can also use mobility to seek new solidarities, experiences, challenges and feelings; sustaining mobility cultures that comprise inter-subjective and co-operative acts that enable individuals to move, communicate and dwell in more dynamic, complex ways.

I examine one alternative way of practicing mobility, a creative bodily practice that seems to work outside regulatory processes and the cultural and technological achievements accelerated by tourist cultures and modern industrial societies. The practice of hitch-hiking, governed through appeals to 'freedom' and an individual desire to inscribe one's own rhythm on the world, offers both geographical movement and a way into an alternative mobility culture that gives value to the turbulence, risk, friction, slower speeds and social exchange it engenders. This chapter gains insights into the practice by understanding hitch-hikers' attitudes towards mobility, describing how their performance of 'motorscapes' creates border-crossing geographies of circulation, multi-locationality and exchange. I argue that their resistant mobilities unsettle the familiar and expected ways of moving, dwelling and doing as they trade speed, convenience and time (rather than cash) for experiences, encounters and connections.

Although a constant in the world of modernity and contemporary modern society, mobility has acquired new dimensions in the late modern (postmodern) context – a constant movement of images, ideas, capital and people (Lash & Urry, 1994). 'Staggering developments in communication and transportation' (Cresswell, 2006: 20) have changed 'our apprehension of space, time and subjectivity' (Simonsen, 2004: 43), creating and stretching new and existing social, cultural and economic networks. In the industrialised West, there have been steady, long-term trends in mobility behaviour across generations in proportion to ascending social mobility. This has heightened the obligations of individuals to be mobile in order to grasp the opportunities that geographic mobility is perceived to give, whether it be in education, leisure or the labour market, unleashing and accelerating various mobilities (Bechmann, 2004). Geographic mobility, although taking place for many reasons, found at many scales and in many forms (migrant, academic, refugee), is acceptable if it is a means to 'get somewhere', encouraged as long as it results in 'improvement' (Cresswell, 1993: 259). However, mobility that is purposeless, without accreditation and aimless is often discouraged and penalised (Stephenson, 2006) while more modern mobilities of circulation replace older ones. Touristic mobilities, related to cultural norms and a developing global tourist culture are continually enhanced, improved and accelerated, enabling them to increase in volume and scope (Urry, 2007), creating a bridge to sightseeing, hotels, rental cars and other shared ways of doing; 'a sub-set of behavioural patterns and values that tend to emerge only when the visitors are travelling' (Williams, 1998: 157).

Given the modern subject's search for authenticity, which Oakes (2006) argues, can only be fleetingly satisfied by a self-induced and self-controlled mobility, tourists routinely display a conscious departure from mobility patterns away from home to engage in global generic mobility codes (hotels, rental cars, flights, restaurants), creating global circulation patterns dominated by aeromobilities (Cwerner et al., 2009) and automobilities (Featherstone et al., 2005). Movement is largely geared towards overcoming the 'friction of distance' when getting from A to B, from home to the destination, while alternative mobility modes are only selected as the tourists engage in expressions of local flavour, norms and customs (such as riding a camel in Egypt, a rickshaw in India, a tuk-tuk in Thailand).

While tourists are not simply cogs in a machine, a tourist culture is learned early in life and performed almost unthinkingly as routine travel and tourist practices, since these practices are based on ability and willingness to habitually use established and approved mobility systems such as aeromobility. However, the individual tourist, rather than being simply a statistical unit, can no longer be confined within a single category, since as a subject and an agent in their own right, their agency over time-space means that they are moving in more complex and dynamic ways. Cresswell (2006: 45) argues '[n]ot only does the world appear to be more mobile, but our ways of knowing the world have also become more fluid' which possibly might not just change the world but our ways of knowing it. Thus, mobility-related practices can sometimes produce social identification and meanings beyond state-led mobility politics, family units and touristic discourses. Innovations in communications, transport and information technology can lead 'to new ways of seeing self and other, places and territories, and ultimately the social and material environment of the contemporary world' (Jensen, 2009: xv). Whether it is travelling the country in a Recreational Vehicle (RV), parkour in an urban centre or overlanding the Maghreb on a trail bike, issues of 'movement, of too little movement or too much, or of the wrong sort or at the wrong time' (Sheller & Urry, 2006: 208) are increasingly of interest to researchers seeking out mobile, lived experiences.

Alternative Mobility Cultures

From increases in car ownership, private motorised transport to out-bound leisure travel, individuals in post-industrial democratic societies are exposed to a range of possibilities to relocate themselves in social and geographical space; the type, range and scope of physical movement driving lifestyles and changes in peoples' lives. Most individuals will, by and large, adopt existing mobility modes upon reaching adulthood, thus ensuring that these developments continue and intensify across generations. From buying the first car or taking the first short-break holiday, institutionalised mobility practices have become a rite-of-passage for many, and it is these practices that offer the means for achieving structured-ness and orderliness as well as social and economic inclusion. Based on norms and regulatory discourses that generate expectations and frames of reference that are used to interpret the behaviour of others, individuals with the perquisite mobility capital can embark on a tourism career that is enabled by a multi-faceted tourism industry. From tour operators, online booking portals, price comparison sites, budget airlines to hotels, the expansion of modern tourism and travel now seems more directly linked to institutionalised and accelerating mobility practices, with the industry providing and elaborating experiences that require tourists to trust networks of supporting intermediaries and mobility systems to provide their basic needs. From the airline to the car system, mobility systems organise 'around the processes that circulate people, objects and information at various spatial ranges and speeds' (Urry, 2007: 52). When tourists buy accommodation, flight tickets, food, transport and entertainment, they buy many things – such as time, space, convenience, security, speed – and are embedded in a tourist culture. This culture is founded on habit and a system where trust in strangers is a kind of institutionalised intimacy, reducing each interaction to a series of abstract signs where real relationships do not exist, since the system simply maximises the flow of tourists in relation to transport, accommodation and the destination itself (Dickinson & Lumbsdon, 2010). Self-perpetuating, these systems affect movement, methods of communication and interaction, producing faux hospitality and 'human relations in the service-sector style' (Baudrillard, 1998: 162) by establishing relationships between elements that would otherwise have no connection.

This individualism can also enable individuals, through these systems of mobility, to find agency, where deliberate choices can construct experiences. How we move becomes 'a statement of class, identity, personality, environmental values and wealth (amongst other things), as well as a practical means of simply completing everyday tasks' (Pooley et al., 2005: 14–15). Whether it is daily, temporary or long-term movement, choice of mobility mode and ways of moving (walking, skateboarding, cycling, car, train), can have implications for social interaction and the negotiation of identities and lifestyle, since each practice has normative regulating principles, norms and socialisation processes (Jensen, 2009). Such practices can become signifying practices and become embodied within alternative mobility cultures that are not derivative of any particular nation-state, migratory network or tourist culture. Such cultural communities are 'often established by people together tackling the world around them with familiar manoeuvres' (Frykman & Löfgren, 1996: 10–11).

From sailing, inter-railing, urban-exploration, skateboarding to motor biking, individuals are increasingly seeking to experience different scales of mobility and dwelling – breaking with everyday life through 'marginal mobility practices' (Jensen, 2009: xvii). This can generate 'a communal way of seeing the world in consistent terms, sharing a host of reference points which provide the basis for shared discursive and practical habits' (Edensor, 2010: 8). Particular practices such as cycling, sailing and walking have become associated with 'slow' travel and tourism (Dickinson & Lumbsdon, 2010) since they give themselves easily to the argument that slow mobilities equate with quality time, physically slowing down, ecology, quality experiences, meaning and engagement. This chapter seeks to explore the concept of slow mobilities by looking at the lived experience of hitch-hikers, whose alternative way of practicing mobility is a signifying but marginal practice that is 'performative' (Goffman, 1990) in nature.

Methodology

While the system of 'automobility' and the privileged position of the car/driver assemblage have become the subject of detailed analysis (Featherstone et al., 2005), Sheller and Urry (2006: 209) argue the 'sites, places, and materialities to the mobilities that are already coursing through them' should also be investigated, since individuals 'repeatedly couple and uncouple their paths with other people's paths, institutions, technologies and physical surroundings' (Mels, 2004: 16). Ferguson (2008), for example, analysed the mobile, lived experience of welfare work and the centrality of the car to that work, while Laurier (2004) argued that new technologies enable the car to be re-assembled as an office. Similarly, hitch-hiking, while dependent on cars and other vehicles and their drivers, is not simply about individuals connecting to a form of transport to get from A to B. While hitch-hiking has re-emerged in some countries because of social, economic and political inequality, this chapter looks at those travelling to and participating in 'organised' hitch-hiker gatherings. Utilising a mixed methodological approach, 10 semi-structured interviews with hitch-hikers shape the investigation into what hitch-hikers actually do and how they learn and perform road infrastructure. The interviews (lasting between 25 and 60 minutes) were recorded via Skype in the days after two hitch-hiking 'gatherings' that the author participated in during the summer of 2010. These hitch-hiking gatherings are built upon the fraternity amongst hitch-hikers and usually require participants to hitch-hike to particular destinations to camp and share activities and stories before hitch-hiking home.

I participated in the 3rd German 'Abgefahren' (http://race.abgefahren-ev.de/⸮) gathering by hitch-hiking from Augsburg to Bled in Slovenia (450 km) and camping there with other participants from 4–7 June 2010, before hitch-hiking back to Munich (360 km). I also participated in the 3rd European hitch-hiking week (http://hitchgathering.org/) from 6–9 June 2010 in Sines, Portugal, which required over 2500 km[1] of hitching over six days. While the former event centred on hitch-hiking to Augsburg before announcing a secret destination and creating two-person teams, the Sines event saw participants organise various pre-gatherings across Europe, before a final push towards Portugal so as to arrive for 6 June.

The events make visible the social phenomenon of hitch-hiking, with my participation not only helping to focus on the practice at the level of individual practitioner but also as a means into an alternative mobility culture. As in other fieldwork situations, numerous informal discussions, my own experiences, still images and observations added value to the research – a movement-driven research methodology sensitive to those on the move, where 'being there', means thinking, feeling and performing their world. The work presented in this chapter is part of a larger study on nomadic lifestyles, and the aim here is not to report the findings of this larger study in a systematic way, but instead to use interviews with Adam (male, 24) and Peter (male, 27), who hitch-hike regularly throughout the year, in order to illuminate key aspects of the practice.

Hitch-hiking and Hitch-hikers

Hitch-hiking (thumbing, autostop, *trampen*) is a means of transportation that requires hitch-hikers to ask people, usually drivers who are strangers to them, for a free ride in their automobile or other road vehicle (air, sea and rail hitching also exist) to travel over distance. Over long journeys, the hitch-hiker is usually required to make a variety of connections across time-space, rebelling against the sequestered spatial logic of the motorway and the car-driver assemblage to accumulate numerous 'lifts'. The practice emerged in the 1920s with the advent of the car (Chesters & Smith, 2001), but came to have a social, political and collective value and affect in the 1960s. The practice became a reference point that provided the basis for shared discursive and practical habits, knitting geographically dispersed people together in a generational phenomenon we call the counter-culture. Prior to the 1960s, '[h]itchhiking did not make one a hitchhiker; practice did not lead to iden-tity' (Packer, 2008: 81), but as obligations to custom and tradition broke down, the practice provided new avenues for people to be part of a genera-tional movement unmediated by the family, the community or economic trade relations. An alternative form of cultural participation associated with freedom, anti-establishment nomadism, adventure, escape and discovery established hitch-hiking as normal, acceptable and even pleasurable (Chesters & Smith, 2001), helping to circulate feelings of belonging to an alternative life trajectory across time-space. By the early 1970s however, insurance com-pany publicity, a police crackdown on the practice in the United States, widespread anti-hitch-hiking legislation across the world, and a sensational-ist media had turned hitch-hiking into a feared form of mobility and the hitch-hiker into a sinister figure (Packer, 2008). A generation of drivers and hitch-hikers who had regarded hitch-hiking as acceptable was diminished and hitch-hiking ceased to be a popular, accessible practice; its demise paral-leling the demise of the counter-culture as a whole.

However, the practice had remained widespread in the former Union of Soviet Socialist Republics (USSR) and the past decade has seen hitch-hiking undergo resurgence, its spatialities again having implications for people's identities. Larger numbers are choosing the practice to sustain and build con-nections across cultural, social and geographic boundaries, demonstrating dispositions they already feel they have about themselves. Part of an alterna-tive 'scene' (Irwin, 1973) since the 1960s, hitch-hiking is again taking centre stage as part of a broader nomadic existence that promotes sustainable living, travelling and hospitality. From festivals (for example, the Road Junky Festival 2010), autobiographical travel books (*Riding with Strangers* by Elijah Wald), music (*Europe* by Ghost Mice), documentaries (*Paris 888*, produced and directed by Fabrice Renucci), films (*Into the Wild*, directed by Sean Penn), fic-tion (*Norton's Ghost* by R. Canepa), online magazines (randomroads.org), dis-cussion groups and portals (hitchwiki.org), the social, political, environmental and economic worth of the practice is being reinforced and reaffirmed. Knowledge about the practice, that is how and where to perform it, is built up on websites such as hitchbase.com, digihitch.com, hitchwiki.org and the hitch-hiking forum on couchsurfing.com,[2] all of which promote the practice as a 'symbol of solidarity' (Hebdige, 1976: 93). Such knowledge, once shared with other hitch-hikers online, or offline at gatherings, enables hitch-hikers to become active participants and creators by mapping the world in a particular way, from tagging and sharing hitch-hiking points to road etiquette and promotion. Adam (one of my interviewees) utilised books such as *Into the Wild* by Jon Krakauer (2007), and a number of websites, to build knowledge about the practice, but it was another hitch-hiker who gave him confidence and new insight into the practice.

> I hitchhiked my first day alone. The second day I met a French guy and I travelled with him for two days and he taught me everything about hitchhiking, so I learned from another hitchhiker. (Adam)

All 10 participants interviewed were already knowledgeable about the websites promoting the practice and most had taken up 'interstitial practices' that enabled them to build a viable existence across national boundaries for both short and extended periods; their ways of moving, dwelling and communicating highlighting a particular homophilly. These 'interstitial practices' include hospitality exchange[3] (i.e. couchsurfing.com), dumpster diving, veganism and wild camping, their involvement in one practice affecting access to knowledge about others. The same communal principles, systems of exchange, denial of risk, trust and use of spare resources (without financial payment) are inherent to the various practices, with hospitality exchange and hitchhiking also demonstrating the embodiment of cosmopolitan dispositions,[4] such as extensive mobility, connecting to many places, peoples and cultures and being open to other peoples and cultures (Szerszynski & Urry, 2006).

> It's no coincidence that a lot of couchsurfers are hitchhikers as well. I am sure they have a lot in common. It's the same concept: sharing rides or sharing beds.... In both cases, it's the open-minded people. (Peter)

Performing Motorscapes

While all roads have mobility capital for hitch-hikers, since, as mobility systems, they are used as a mode of circulation (Urry, 2007), various types of roads have different mobility capital depending on the hitch-hiker. For the two gatherings, most participants utilised Europe's motorway network, since it is more organised for circulation and provides quicker access to the two gathering locations. While roads have produced many bodies (horse rider, walker, cyclist), automobility generated new motorways that structured the flow of people and goods along particular routes, linking automobility with progress, modern living, acceleration and freedom, while at the same time eliminating many other forms of mobility types from the roads.

The routinised acceptance of the car and the habitual performance of driving mean that automobility has entered the social and economic fabric of people's lives, a dependence that demands ever-increasing resources. Motorscapes facilitate 'practice but are also reproduced by the actions and understandings of people' (Edensor, 2004: 111). Taken for granted in the developed world, and treated as one of life's constants, everyday and life projects are planned with the assumption that a car will be part of the environment, with most presupposing its continuation as the most feasible mode of ordered mobility. A 'mode of mobility is neither socially necessary nor inevitable yet it seems almost impossible to break away from' (Urry, 2008: 344).

However, motorscapes do not generate one specific kind of body, but instead the roads, its drivers and supporting infrastructure can enable individuals to construct a range of identity positions by negotiating new meanings. Edensor (2006: 385) notes how 'performances are increasingly acted out by competing actors on the same stage', with motorways containing everyone from tourists, businessmen, truck drivers to police officers, some of whom seem appropriate to this domain, and others who have contested ideas about what practices are appropriate. While car-dependent practices are promoted and motorways and infrastructure built, the motorscapes and their amenities can become a territory in which individuals use their (mobile) bodies and senses in 'doing' movement differently. The practice of hitch-hiking requires them to leave their homes to occupy a highly visible space on the world's motorways and roads. While drivers in their 'portable territory' move between fixed points and regular routes, for those hitch-hiking the road offers possibilities and multiplicities in journeys of fluid vectors and changeable paths.

> You meet so many people.... You don't know what time you arrive and where ... to see where the road will take you. (Peter)

> It's all about being on the road, traveling, talking to people I don't know, to go to foreign countries, talk to foreign people ... you get closer to people. (Adam)

More than just a bodily practice, the individual also carries the (cosmopolitan) values, reasons and objects that make the practice work, their very performance of hitch-hiking making motorscapes a fluid social space. An assemblage of signified features, objects, amenities, other bodies and infrastructure is positioned according to the components' ability to meet the requirements of the practice. Service stations become a place where hitch-hikers can replenish their bodies, interact with drivers and sleep overnight, while sprawling transportation networks become a place of movement, encounter and exchange. These motorways and roads constitute their territory with drivers, cars, petrol stations and trucks enabling hitch-hikers to orchestrate in complex and heterogeneous ways their mobilities and socialities, and therefore their identities across very substantial distances. From visiting family and friends to hitch-hiking to the far corners of the globe, spatial mobility becomes both 'the prerequisite for, and the consequence of, social interaction' (Beckmann, 2001: 597).

In order to be able to stay mobile, it is necessary for hitch-hikers to develop ties with drivers and service station workers at varying times and places, a practice that demands human interaction, intimate engagement, exchange and a lot of patience. This is not a one-sided exchange, since those who pick up hitch-hikers usually have excess capacity and giving someone a ride has advantages, providing both driver and hitch-hiker with a 'comparative advantage' (Ricardo, 1817) in trade. Nevertheless, driving (especially on motorways) is based on an ensemble of institutionally defined and socially recognised behaviours based on prescribed ways and rules (traffic systems, speeds, tolls). Some drivers who seek to defend their personal space define any practice that lies outside the range of prescribed ways, irrespective of their potential value, as inappropriate. Many regard hitch-hikers as unsettled, transitional and non-integrated, their mobility deviant, suspect and potentially dangerous. For other drivers, their 'quasi-private' mobility is associated with sexual desire and masculinity (Urry, 2000), with hitch-hiker bodies perceived to be exploitable and easily harassed. However, for those who overcome the perceptions of 'risky strangers' (Furedi, 1997), the hitch-hiker may enliven their mobility experience. While some of the drivers have a sense of obligation and reciprocity, others come from regions of the world where hitch-hiking is more acceptable, or have grown up during a period when it was acceptable.

Drivers recognise the mobility capital in this form of circulation, their willingness to trust outweighing any perception of risk and their representation as a risk group. For both the hitch-hiker and driver, their relationship is based on mutual understanding, the practice requiring a sense of solidarity, a willingness to share and co-operate as well as an active acceptance of difference that is not based on assimilation or imposition. Each lift is categorised as an 'authentic' moment that breaks through 'the dulling monotony of the "taken for granted"' (Shields, 1999: 58, citing Harvey, 1991: 429), with each boundless journey providing incident, new knowledge and insight into lives that are tangible and memorable. The ability to engage equally in exchange relations is based less on binding convention and monetary exchange and more on bodies moving and dwelling together for short but intimate periods to create bonds that are both imagined and felt.

> One guy took us to Sarajevo. And he was really funny. It took us four hours. In the end, we exchanged phone numbers and we went out with him. He invited us for beer and stuff. It was a lot of fun. (Peter)

> They tell you everything about their problems ... they tell you almost everything. (Adam)

The World of Difference

Rather than associate hitch-hiking with the 'slow movement' or disrupted mobilities, respondents believe the practice enhances their lives, enabling them to 'revisit from time to time selective assumptions and dispositions that have gripped us and to refresh our energies to reenter the rat race' (Connolly, 2002: 144). While slow mobilities might be about authenticity, variety, conviviality and the value of travelling, it need not be about relaxation, rest and travelling slowly (Germann Molz, 2009). Considering the turbulence and friction inherent in the practice (i.e. waiting for lifts, sleeping at motorway service stations, harassment, exposure to the automobile driver's gaze), participants talk about their journeys (encounters, experiences) as something valuable and spatially inspiring, with the potential to affect, develop and transform the self. The control they feel they have over their mobility provides them with an immense source of pride. When compared with the 'seamlessness' of car journeys (since you do not have to change modes of travel to get from A to B), hitch-hiking as a mode of self-directed mobility may be highly fragmented and inconvenient, bringing turbulence, friction and slower speeds. However, the intensities of encounters mean the travellers remain open and sensitive to the affects outside the self, where sensations and physical efforts are acutely felt.

The practice is both constructive and unsettling, with each journey associated with complex feelings, unexpected moments, instant decisions, gratitude, fear, contemplation, excitement, boredom, marginalisation, and where even risk is experienced as liberation, helping them to recover 'the world of difference – the natural, the sensory/sensual, sexuality and pleasure' (Lefebvre, 1991: 50). Rather than seeing the practice as slow or fast, the hitch-hikers argue the real world is navigated and browsed more efficiently with their bodies, their movement more 'real' than those that could be accomplished by the passive and risk averse tourist. Those without the time, stamina or bodily/physical competences or skills cannot travel the 'right way' and cannot actualise the possibilities that the road truly offers like the hitch-hiker.

While a minor and marginal mode of moving and dwelling, the practice is a skilful one, with Kesselring and Volg (2008: 167) suggesting that 'mobility' can be defined 'as an actor's competence to realize specific projects and plans while being on the move'. Taking a grip of the speed that surrounds them, hitch-hikers believe their mobility choice enables them to eradicate the boundaries between public-private, insider-outsider, mobility-immobility, virtual-physical, host-guest, traveller-tourist and us-them, the dichotomies of fast-slow not accounting for the complex interplay between patterns of mobility that bifurcate within a mobility continuum. By appropriating, respecting and transgressing the motorscapes, the hitch-hikers believe their mobility illuminates events, peoples and places that a guidebook could not, telling 'spatial stories that cannot be read' (Tonkiss, 2003: 304).

The distinct forms of sociability and encounters with drivers, long wait-ing times, places visited and hospitality exchanged create the shared experi-ences of movement which are central to the physical gatherings and online discussion boards. The practice is circulated through the online discussion boards, books and event spaces advertised on hitch-hiking discussion boards, the synchronisation of manoeuvres as people hitch-hike to various spaces becoming part of a larger performance; one that allows participants to con-struct and communicate their belonging to an alternative life trajectory. The hitch-hiker gatherings, growing across Europe and North America, are based on the premise that the more people who hitch-hike, the more likely par-ents, educators, authorities, the media and drivers can overcome potential psychological barriers and social distrust, since fear once visited on the prac-tice is difficult to eliminate. As well as enabling first-time hitch-hikers to share risk by hitch-hiking with experienced others, communal eating, shar-ing information, hitch-hiking stories and identifying areas of agreement are central to the gatherings. While most of the participants undertake the practice by way of episodic mobility rather than nomadically, hitch-hiking to the gatherings demonstrates both individual skill and attempts to cross over to a more alternative life trajectory; their take up of the practice con-stitutive of a transformative way of thinking about moving, dwelling and communicating.

> I did not know any hitchhikers before. I just want to meet other people who love hitchhiking. (Adam)

Experiences are decoded during informal group discussion, providing a plat-form for recognition, validation and socialisation, while individual experi-ences (often blogged) are used as a set of resources to assert one's self and, in doing so, demonstrate belonging. Roads, their inhabitants, materialities, infrastructure, as well as the opportunities, means and constraints they offer, produce a sense of familiar space for hitch-hikers since they negotiate them through the same embodied practices.

'Resistant' Mobilities

I argue that one can no longer speak of a hegemonic tourist culture, since attachment and identity are negotiated through self-induced and controlled mobility. The material productions of different mobilities, some of which emerge as self-sustaining mobility cultures, remain under-researched and unsupported. Rather than accept and map various practices associated with alternative ways of moving and dwelling, the tourism industry, grounded in control, can often be hostile to such practices. This is especially so if such practices are seen to constitute or represent resistant mobilities, insofar such practices may account for individuals moving away from tourist practices enabled through intermediaries.

Similar to those involved in the Slow City or Slow Food movements, rather than resistant to speed, I argue hitch-hikers are resistant to 'placelessness' and the loss of human interaction; alternative mobilities highlighting the hidden power (away from the motel/hotel circuits, tourist offices) of the local in a global borderless world. Practices with 'different pacings and pulses which critique normative, disciplinary rhythms and offer unconventional, sometimes utopian visions of different temporalities' (Edensor, 2010: 16) are emerging from the bottom up, but are often contested. I am not arguing that a practice like hitch-hiking or other marginal and liminal mobilities such as walking, wild camping or cycling reverses the tourist gaze, unmasks the tourism industry and lays the seeds of destruction for institutionalised mobility practices, but like the slow movement, resistance to quick, solitary consumption is emerging. Whereas motorscapes have come to be associated with automobility, hitch-hiking, a practice once undermined, is again being sought by individuals seeking to cross cultural, social and physical boundaries for genuine social connectedness; resistant to the imposition of accelerating rhythms and normative patterns of movement. While it is not part of a 'slow movement', the practice has some, if not all, of the slow movement's attributes, as hitch-hikers appreciate the distances they travel, the various drivers and the time it takes to travel.

> I learned a lot about myself. I would have never thought when I wait for twenty hours in one spot that I am such a calm, patient guy. I learned a lot about people from foreign countries. It makes you more open-minded. (Adam)

> You should be friendly and very, very patient. If you stay at one place for two, three or four hours; you have to be patient. (Adam)

Conclusion

Worldwide communications, trade and technologies have given a new impetus to human mobility. Contemporary Western societies are increasingly organised around movement, where innovations in transportation and communications form the backbone of a more accelerating, interconnected world. Machine-based fast movement and automobility in particular have dominated in an era of globalisation, becoming a seamless and feasible mode of everyday life transport for many, sustaining networked patterns of social and economic life. However, changing mobility preferences mean that various ways of dwelling, communicating and moving have (re)emerged in relation not only to economic strife and/or alienation, but also because of environmental pressures and political and technological changes. Self-induced mobility, when incorporated and appropriated as a practice, can enable new multiplicities, possibilities and openings for individuals. New learnt norms, rituals, tempos and rhythms can constitute an important part of group identity as people come to share understanding through various ways of moving, reinforced by the monitoring of self and others through bodily techniques and dispositions. While habitual tourist practices are well researched, non-institutionalised practices that are neither neat instances of smooth, seamless and ordered mobility based on a bodily practice which are contributing to the shaping of contemporary practices of consumption, production and lifestyle, are under researched. This is because they are often minor practices, few are willing to 'learn bodily' and put their body on the line when more convenient and speedier modes of dwelling, eating and moving are available.

Often derided as being a minor, episodic and occasional mobility practice, this chapter has taken a fresh look at hitch-hiking, which has again developed into part of an alternative mobility culture practiced by individuals searching for and/or demonstrating an alternative life trajectory, their alternative mobilities a function of movement within accelerating motorscapes. Rather than feeling disempowered or powerless in an accelerating world, this chapter argued that hitch-hikers greatly value mobility as a means of bringing about connections. Participation in this mobility culture demands certain cosmopolitan dispositions and a corporeal engagement that necessitates the mastering of the body. It is a corporeal technique that enables those who practice it to live, even temporarily, a marginal life outside of the expectations; economic, political, behavioural codes; constituencies of interest and the accelerating mobilities that accompany modern societies. This practice, and others such as wild camping, parkour, nomad houses and hospitality exchange, create new opportunities and risks for regulatory authorities and tourism planners. Rather than advocate for policy interventions that facilitate alternative mobility cultures, the tourism industry routinely looks for interventions to eliminate them, since such alternative mobility cultures are unlikely to trade cash for speed, security, safety and convenience. However, such practices will continue to emerge and spread as people enact and perform movement in new ways, reasserting and communicating new corporeal claims to space through embodied practices.

Notes

(1) I hitch-hiked from Dieppe, France after catching a night ferry from Newhaven, United Kingdom. From there, I hitch-hiked to Barcelona for a 'pre-gathering' before hitch-hiking to Sines. I primarily slept close to service stations at night. Once there, I camped with other participants for three days.

(2) By 16 July 2010, Hitchwiki had 1600 articles with a total of 134,583 edits and an average of 1300 visitors per day. Couchsurfing.com, a hospitality exchange, site has a hitch-hiking discussion board with 15,000 members.

(3) Hospitalityclub.org founded in 2000, the first database-driven hospitality exchange site, was based in part on the 'Hitch-hiker's Registry' (2000), a hitch-hiker online accommodation network programmed by Bernd Wechner after the initial idea by Vilnius Hitch-hiking Club members, Vladas Sapranavicius and Augustas Kligys who set up an email-based hitch-hikers' accommodation list.

(4) While it is possible to hitch-hike without cosmopolitan dispositions (since they are not universal dispositions), participants at the gathering and on the online discussion boards communicate and demonstrate these dispositions. Their movement to the gatherings and the gatherings themselves crosscut social, political, economic and cultural boundaries and demonstrate a cosmopolitan Europe.

References

Baudrillard, J (1998) *The Consumer Society: Myths and Structures*. London: SAGE Publications.

Bechmann, J. (2004) Ambivalent spaces of restlessness: Ordering (im)mobilities at airports. In J.O. Bærenholdt and K. Simonsen (eds) *Space Odysseys: Spatiality and Social Relations in the 21st Century* (pp. 27–42). Aldershot: Ashgate.

Beckmann J, (2001) Automobility – A social problem and theoretical concept. *Environment and Planning D: Society and Space* 19 (5), 593–607.

Cancpa, R. (2010) *Norton's Ghost*. Gainesville, FL: Skinny Wizard Media.

Chesters, G. and Smith, D. (2001) The neglected art of hitch-hiking: Risk, trust and sustainability. *Sociological Research Online*, 6 (3) – Online document: http://www.socresonline.org.uk/6/3/chesters.html

Connolly, W.E. (2002) *Neuropolitics: Thinking, Culture, Speed*. Minneapolis, Minnesota: University of Minnesota Press.

Cresswell, T. (1993) Mobility as resistance: A geographical reading of Kerouac's 'On the Road'. *Transactions, Institute of British Geographers* 18, 249–262.

Cresswell, T. (2006) *On the Move: Mobility in the Modern Western World*. London: Routledge.

Cwerner, S., Kesselring, S. and Urry, J. (eds) (2009) *Aeromobilities*. London: Routledge.

Dickinson, J. and Lumbsdon, L. (2010) *Slow Travel and Tourism*. London: Earthscan.

Edensor, T. (2004) Automobility and national identity: Representation, geography and driving practice. *Theory, Culture & Society* 21 (4–5), 101–120.

Edensor, T. (2006) Performing rurality. In P. Cloke, T. Marsden and P.H. Mooney (eds) *Handbook of Rural Studies* (pp. 484–495). London: SAGE Publications.

Edensor, T. (2010) Introduction: Thinking about rhythm and space. In T. Edensor (ed.) *Geographies of Rhythm: Nature, Place, Mobilities and Bodies* (pp. 1–18). Aldershot: Ashgate.

Featherstone, M., Thrift, N. and Urry, J. (eds) (2005) *Automobilities*. London: SAGE Publications.

Ferguson, H. (2008) Liquid social work: Welfare interventions as mobile practices. *British Journal of Social Work* 38 (3), 561–579.

Frykman, J. and Löfgren, O. (1996) Introduction: The study of Swedish customs and habits. In J. Frykman and O. Löfgren (eds) *Forces of Habit: Exploring Everyday Culture* (pp. 5–19). Lund: Lund University Press.

Furedi, F. (1997) *Culture of Fear: Risk-Taking and the Morality of Low Expectation*. London: Cassell.

Germann Molz, J. (2009) Representing pace in tourism mobilities: Staycations, Slow Travel and The Amazing Race. *Journal of Tourism and Cultural Change* 7 (4), 270–286.

Goffman, E. (1990) *The Presentation of Self in Everyday Life*. London: Penguin.

Hebdige, D. (1976) *Subculture: The Meaning of Style*. Methuen: London.

Irwin, J. (1973) Surfing: The natural history of an urban scene. *Urban Life and Culture* 2 (2), 131–160.

Jensen, O.B. (2009) Foreword: Mobilities as culture. In P. Vannini (ed.) *The Cultures of Alternative Mobilities: Routes Less Travelled* (pp. xv–xix). Farnham: Ashgate.

Kesselring, S. and Volg, G. (2008) Networks, scapes and flows – Mobility pioneers between first and second modernity. In W. Canzler, V. Kaufmann and S. Kesselring (eds) *Tracing Mobilities* (pp. 163–180). Aldershot: Ashgate.

Krakauer, J. (1996) *Into the Wild*. New York: Villard Books.

Lash, S. and Urry, J. (1994) *Economies of Signs and Space (Theory, Culture & Society)*. London: Sage.

Laurier, E. (2004) Doing office work on the motorway. *Theory Culture and Society* 21 (4–5), 261–277.

Lefebvre, H. (1991) *The Production of Space*. Oxford: Blackwell.

Mels, T. (2004) Introduction: Lineages of a geography of rhythms. In T. Mels (ed.) *Re-animating Place: A Geography of Rhythms* (pp. 3–44). Aldershot: Ashgate.

Oakes, T. (2006) Tourism and the modern subject: Placing the encounter between tourist and other. In C. Cartier and A. Lew (eds) *Seductions of Place: Geographical Perspectives on Globalization and Touristed Landscapes* (pp. 36–55). London: Routledge.

Packer, J. (2008) *Mobility without Mayhem: Safety, Cars, Citizenship*. Durham, NC: Duke University Press.

Pooley, C., Turnbull, J. and Adams, M. (2005) *A Mobile Century? Changes in Everyday Mobility in Britain in the Twentieth Century*. Aldershot: Ashgate.

Ricardo, D. (1817) *On the Principles of Political Economy and Taxation*. London: John Murray.

Sheller, M. and Urry, J. (2006) The new mobilities paradigm. *Environment and Planning A* 38 (2), 207–226.

Shields, R. (1999) *Lefebvre, Love & Struggle*. London: Routledge.

Simonsen, K. (2004) Spatiality, temporality and the construction of the city. In J.O. Bærenholdt and K. Simonsen (eds) *Space Odysseys: Spatiality and Social Relation in the 21st Century* (pp. 43–61). Aldershot: Ashgate.

Stephenson, S. (2006) *Crossing the Line. Vagrancy, Homelessness and Social Displacement in Russia*. Aldershot: Ashgate.

Szerszynski, B. and Urry, J. (2006) Visuality, mobility, and the cosmopolitan: Inhabiting the world from afar. *British Journal of Sociology* 57 (1), 113–131.

Tonkiss, F. (2003) The ethics of indifference: Community and solitude in the city. *International Journal of Cultural Studies* 6 (3), 297–311.

Urry, J. (2000) *Sociology Beyond Societies*. London: Routledge.

Urry, J. (2007) *Mobilities*. Cambridge: Polity Press.

Urry, J. (2008) Governance, flows, and the end of the car system? *Global Environmental Change* 18 (3), 343–349.

Wald, E. (2006) *Riding with Strangers: A Hitchhiker's Journey*. Chicago: Chicago Review Press.

Williams, S. (1998) *Tourism Geography*. London: Routledge.

11 'If You're Making Waves Then You Have to Slow Down': Slow Tourism and Canals

Julia Fallon

This chapter will examine slow travel and tourism by referring to both the historical and present day canal landscape. Canals comprise a man-made cut into the earth and alongside these there are buildings constructed originally to support freight transport or water supply. These historical linear park areas are serving as leisure space for boating and walking where it is possible to slow down and escape from the everyday. Slow tourism and canals fit because in this context the principles of slow travel and tourism are manifest. These principles are: spending quality time, experiencing a place at a pace allowing meaning and engagement plus being in tune with ecology and diversity (Dickinson & Lumsdon, 2010). Canals, it is argued here, especially the narrow ones in Europe restricted by limited space, encourage a slow tourism experience. Slow tourists are in search of a relaxing, passive, traditional experience with low technology and a lack of competitiveness as an antidote to a world where the sound-bite features significantly over substance (Palmer, 2002).

The Canal Context

The building of the canals and development of their locations made distinct changes to the landscape (often over considerable time periods), introducing water where it had not been before and physically changing the environmental systems and ecology (Hadfield, 1986). In places like Exeter, UK, the canal became a significant attraction and a place that offered 'reflective space' (Tresidder, 2000: 303) a location away from it all and one where time was forgotten if only temporarily. Hoskins' description of the Exeter canal as offering a favourite walk elucidates this landscape of the canal location as one that was:

Peaceful winding through the meadows of the Exe valley past congenial inns ... in the pastoral settings the canal followed clear and sparkling in the sunshine something new in the landscape with their towpaths, lock-keepers cottages, stables for canal-horses, their navigation or canal inns where they met a main road and their long and gaily painted boats. (Hoskins, 1981: 253)

As Hoskins (1981: 247) explains of the English 18th-century experience, not only 'did they bring stretches of water to areas where there was previously none with the changes in bird and plant-life' they aroused a curiosity in a man-made environment, and goes on to say that: 'For the first time there were aqueducts, cuttings and embankments, tunnels locks lifts and inclined planes and many attractive bridges and they greatly influenced the growth and appearance of many towns' (Hoskins, 1981: 248).

For economic reasons, canal construction in the past made very good sense, but it was often pragmatic, meeting local needs rather than being part of an overall transport strategy. The European canals are, with the exception of the Swedish Kiel Canal, narrow and shallow and their role in providing transportation for freight has been superseded by leisure use. The Gota Canal, which links the Swedish west coast with Stockholm's environs, for example, was completed in 1832 but then left to decline because the railway offered faster and cheaper services. The canal is now a popular yachting stretch every summer (Heidbrink, 2010) and the reason for this popularity, apart from the pretty location, is that amateurs and learners enjoy this type of water for its slower pace and the opportunity to enjoy their surroundings. Canals therefore, offer the experience of a 'slow-moving form of travelling' (Erfurt-Cooper, 2009: 104) where the boat can be navigated whilst allowing the experience of the place. This type of holiday has become very popular in France where barges are used on rivers and canals and holidaymakers are encouraged to moor alongside the canal and spend time cycling around towns at a pace that fosters awareness of local distinctiveness (Erfurt-Cooper, 2009). Such holiday experiences are characterised by shorter distances, lower carbon consumption with a greater emphasis on the mode of transport, which also encourages place-based knowledge in what Dickinson and Lumsdon (2010: 10) recognise as the 'slow travel mindset'.

As places to visit or travel upon, canals have been an emotive topic for many members of the public interested in the past and its preservation. The movement to restore canals includes many volunteers that have both campaigned and worked hard physically in clearing up/maintaining/restoring. In researching slow tourism in the context of canals this therefore raises issues about the work/leisure overlap, when there are so many seeking a different experience by volunteering in making a contribution to restoration and renewal. The author has already explored this topic elsewhere (see Trapp-Fallon, 2007) and discovered that the enthusiasm about the preservation of the past by volunteers has shaped the slow travel and tourism experience today.

Canals and Slow

There is probably little recognition that this type of slow tourism experience dates from the early 20th century when the National Inland Navigation League, founded in 1919, did much to revive interest in the canals in the 1920s (Squires, 1979: 19). A clutch of books was published in the UK at that time, including *Canals, Cruises and Contentment* by Austin E. Neal in 1921, and these positively encouraged pleasure cruising. The appeal of being on the water in a canal boat was not new even then however, as Charles Dickens (2004: 170) had also described travel on canals in his *American Notes*:

> The lazy motion of the boat, when one lay idly on deck, looking through rather than at, the deep blue sky; the gliding on, at night, so noiselessly, past frowning hills, sullen with dark trees and sometimes angry in one red burning spot high up, where unseen men lay crouching round a fire; the shining out of bright stars, undisturbed by wheels or steam, or any other sound than the rippling of water as the boat went on: all these were pure delights.

Interestingly here, Dickens is writing in response to the alternative forms of transport available at that time. Already the dehumanisation and disengagement from the destination associated with the speed of travel created by steam was prompting a reaction from a tourist and traveller. Dickens had had rather upsetting experiences whilst journeying across the Atlantic, but in the calm canal waters he was able to modify his views and positively experience travel and transport by boat.

This rather pioneering aspect of travelling by water transport for leisure is documented in 19th-century British publications like *The Field* and *The Canoeist*. Vine (1983) suggests that those that wrote about their experiences were, in their way, similar to explorers going to Africa and the Arctic. He goes on to record that during the early 19th century there were a number of travellers who wrote about their experiences travelling on water throughout Europe. There were those that used public transport, like Robert Southey, who wrote about his travels on the Dutch and Belgian canals in 1815, but Vine was most interested in those who travelled independently. These authors were using a variety of different sailing craft and their experience encapsulates escapism, here described by Manfield (cited in Vine, 1983: vvi):

> It is the feeling of perfect independence and freedom in those extensive solitudes, where a human being scarcely ever sets his foot and where the silence was only broken by the dull roaring of a rapid the booming of a the bittern or the rush and rattle and rush of the wings of the wild geese.

What Manfield describes is what Macbeth (2000) believes to be the utopian ideal. Moreover, this links with searches for authenticity; an adventure for personal growth and development often discussed in the motivation for tourism but which Macbeth succinctly sums up as people who are looking for something more and better in their lives, something that enlivens and enriches who they are. This experience, while describing an excitement of travelling in the 19th century, is not so remote now. The attraction of the boats, water, green space or a combination of these has been recorded as a pull factor when considering a destination's attractiveness (Verbeke, 1986, cited in Page, 1994) and is recognition of the human desire to have a connection with nature and water. Slow travel and tourism thinking is a reminder of this and prompts us to ponder on Wordsworth's words (cited in De Botton, 2003: 141) whilst admiring the Wye valley:

> ... [nature] can so inform
> The mind that is within us, so impress
> With quietness and beauty, and so feed
> With lofty thoughts

This view is one that has been written about more recently by authors like Conway and Timms (2010), believing that a remote location and under-developed space nurtures slow tourism, and here they are at one with Wordsworth in thinking that the outdoors, nature and the removal from a hectic schedule allows the rediscovered sense of self (also mentioned by Dickinson and Lumsdon, 2010). They are also progressing the ideas of authors like Hoskins (1955, cited in Matless, 1994: 29) who reflected that traditions offer a 'refuge and a solace to be cherished'.

This is a perspective espoused by Rolt. As an engineer, he was fascinated with canals and wrote at a time that Black and MacRaild (2000: 60) refer to as the 'birth of an obsession with the technology of the Industrial Revolution'. Rolt wrote an account of his trip on a canal boat from Banbury through the Midland canals in 1939, which he saw as a celebration of what Ackroyd (2004: 149) called the 'vernacular landscape'. He wrote of how the canals represented a lost time, a place to reconnect with the past and recreate it. Rolt, as an early restoration pioneer, described as a 'romantic' by fellow canal enthusiast and author Charles Hadfield (1994), would now be held up as an advocate for slow travel and tourism.

Canals and Heritage

There are those that say that Rolt tried to create what had never existed because he sentimentalised life on the canals (Harvie, 2002). It appears that he invented some aspects of the lives of the canal workers and sought to give some legitimacy to his words by the connection of canals with an industrial past (Owen, 1991), but this is not so unusual in tourism where heritage attractions combine history with tourism demand. Rolt's motivation was clear in that he did not wish the industrial past to be forgotten and consequently he campaigned to retain any evidence. This is not so easily done, and as Graham *et al.* have noted, the worth attributed to the canals became less about their intrinsic merit but rather more about a complex array of contemporary values, demands and even moralities (Graham *et al.*, 2000: 14).

This resonates with the present day. Rolt's desire to retain and recreate the past is seen regularly in canal-related events where there is a celebration of the arts, history and food and drink within local communities (Towpath Talk, 2010). Slowness is accentuated, for example, quite literally by horse-drawn boats – themselves an attraction – which represent travelling at the right speed for the experience, changing attitudes to time and having a quality rather than a quantity of experience (Dickinson & Lumsdon, 2010). These local events often attempt to give an impression of the past and are regularly in evidence at museums, festivals and local events and celebrations.

To some extent activities of this kind have been part of the restoration and regeneration activities and a reaction to changes in society. The 1960s was the period when Rolt's views about the canals were more widely accepted and there was a drive 'to preserve for all time a useful amenity for non-motorists to enjoy their leisure away from the danger and noise of the roads' (Inland Waterways Association, 1953). This added to concerns about the role of the car and its influence on the environment. Ruskin (cited in Parkins, 2004: 365) states that there is a 'destruction and dehumanisation' in fast travel, and Dennis and Urry (2009: 133) believe that the car: '... redefined movement, pleasure and emotion ... transforming the fitness of landscape for mobility systems that have to find their place within a landscape formed and maintained by this car system'.

Environmental concerns and a re-evaluation of the travel experience have led to resurgence in interest about water transport, often unpowered or where the distances travelled are so little that the carbon footprint is seen as too limited for major concerns (Dickinson & Lumsdon, 2010). One tale that captures the imagination is for example, Jerome K. Jerome's humorous tale, *Three Men in a Boat, Not to Mention the Dog!* first published in 1889 (which demonstrated the shift to leisure enjoyment in the countryside after urbanisation had changed the way of life for many in the West). This book has had several iterations for television, with re-enactments of experiences of travelling on the River Thames by celebrities experiencing local hostelries and engaging with mooring places and local people. All of this is seen as fun and a pleasurable way of spending leisure time.

These entertainments also reflect something of the academic writing on the tourism experience. For example, Reisinger and Turner (2002) cite Harlak

(1994) in acknowledging the importance of the relationships between local people and tourists, believing these to significantly influence the tourist experience. Also Fagence's (1998: 257) work includes the belief that 'it is the richness of the interaction between residents and tourists which gives the region its particular attractiveness'. Similarly, Alastair Sawday (2011), the author of *Go Slow England: Special Places to Stay, Slow Travel and Slow Food*, advocates this connected experience. Sawday writes about 'slowness' and his opinions endorse the research used by Visit Britain, revealing that visitors/tourists want an authentic experience (Visit Britain, 2011). History and culture are wanted by visitors, described by Carnegie and McCabe (2008: 360) as 'playing out with time'. Canal environs often provide a strong sense of the history of the place visited and offer opportunities to connect with and explore the past (Carnegie & McCabe, 2008).

According to research conducted by Visit Britain (2011), this seeking of the authentic and the real is not only for the domestic market but is also sought by visitors from abroad. The search for the authentic is believed a priority by these visitors, whose desire for immersion in the culture along with a sense of nostalgia is linked with the Slow Food and Slow Cities perspective (Visit Britain, 2011). Sawday with Mckenzie (2008), whilst recognising that visitors need support in reaching the right decisions on where to spend their time and money, has produced publications where he offers ethical collections of tourism suppliers who are judged by their environmental footprint, their role in the local community and their serving of local and organic food.

Sawday (2008) uses the Slow Food movement philosophy and his experience as a volunteer in the developing world (places that are already slow according to Dickinson & Lumsdon, 2010) to inform his choices in suggesting places to visit and stay in England, with an understanding that slow means 'local, grown with respect and integrity and with thought to the consequences' (Sawday, 2008: 2). His philosophy resonates with the sustainability agenda within the tourism literature where there are concerns about the environment, local economies and perpetuating small-scale Indigenous enterprise. Sawday (2008: 2), an advocate of the slow movement which involves a conscious effort to make changes and a redefinition of places, explains: 'Where cities are thoughtful places: that value peace and quiet, local production, people over cars, a dark night sky, high quality artisan production, low-energy consumption and importantly, time to enjoy all these things within a community'.

Whilst not mentioning tourism explicitly, Sawday sees the future of cities as rather like canals, that is, as places where there can be a unique interaction between landscape, local community and tourists. Parkins (2004: 371) similarly draws upon the writing about the Slow Food movement, where there is an emphasis upon connectedness, tranquillity and community, making it relevant to the tourism experience. This is reinforced by the actors and boat owners Prunella Scales and Timothy West (cited in Atterbury, 1994):

There is friendliness, a corporate spirit among canal boaters which is perhaps not characteristic of their counterparts on a river. River people tend to display an appetite for competition, as to size, appearance and speed of their craft. Chugging along a 15ft channel in a narrow boat at 3 and a half knots, however provides little challenge to the competitive spirit.

Haywood (2008: 64), another canal enthusiast, describes the situation in England when he says:

> Every weekend, all over the country, thousands of us flock to our boats, and no sooner do we feel the water under our feet then we turn into totally different creatures. Urban men ... suddenly go all social and walk along the towpath with an inane smile on their face passing the time of day with anyone and anything.

Being on and within the space surrounding canals and being on and near the water within the uncomplicated boundaries of a linear park encourages an attitude, one of being apart from the modern and hectic. The space for reflection, or to experience something of the past, is the antidote for some in coping with the pressures of the time-pressed present. Finding green spaces and park areas is important in our measurement of the quality of our lives and so green spaces surrounding canals are arguably similar to Greenbie's (1981) conception of parks, which he describes as friendly places where it is possible to enjoy being alive and getting close to nature.

The motivations then for seeking experience on canals would be for the sense of being apart, being close to water, nature, seeking a perceived authentic experience and allowing a closeness to the past and a recognition for 'an understanding of a life that is no longer existing' (Halsall, 2001: 159). There is a new space-time relationship in this world and the 'pace of the journey timing is an authenticating tool' (Halsall, 2001: 159). In these canal spaces, often now seen as linear parks, there are opportunities for people to experience time to reflect. As Andrews (2006) indicates, that reflection allows joy because it is only when you have time to think that you can really experience the joy that is being felt. It is also, she continues, the way to be aware of what is happening and what is being felt. Leisure, Andrews goes on to say, is the way that we learn because without asking ourselves why we are doing what we are doing, we have 'a second-hand-life' (Andrews, 2006: 145).

Thus, it can be argued that canal boating says much about the way that people choose to explore/enjoy nature and appreciate the beauty of the outdoors. The size and depth of canals dictates the speed of travel and those that travel upon them. Slowness is synonymous with canals, in fact, canal boaters are seen to be people who have chosen to slow down and one reason for this is that the speed limit for boating is literally four miles per hour in the UK. Similarly in the USA, although speed limits vary from state to state, on the canals they refer to the need for a no-wave zone. According to the UK Good Boating Behaviour Code, 'if you're making waves then you have to slow down'. This speed is equated to a brisk walking pace and boaters are advised that if they are faster than walkers they must reduce their throttle, in the same way that they should halve their speed for passing boaters and anglers (Canal Cuttings, 2011). 'Take your time – it is the essence of the holiday spirit anyway', suggested Neal (1921: 21) almost 100 years ago, but there are also practical reasons for going slow on canals. The space is confined and the water that is moved by the boat cannot go far. If this is pushed ahead and the stern of the boat drops it can get clogged with weeds or collide with rocks in the shallow depths. By opening up the throttle and creating waves, the canal banks erode and the strain for speed could cause engine trouble. Neal describes this as a sort of 'poetic justice' (1921: 21).

The relevance of these observations would be acknowledged by those using canals today. This is surprising since Neal was writing in 1921 and the canals in the UK in the interim period have experienced decline and subsequent restoration/regeneration. He would no doubt be pleased to know about the formal recognition of canals and their more recently acknowledged role in tourism and leisure supply (e.g. see IWA and British Waterways websites). This investment has been informed partly by further examination of canals from a tourism and leisure perspective, which shows that in research into spaces where there is water, the water is revealed as the primary aspect of their attractiveness.

Furthermore, this was recognised over 30 years ago by the author Jansen-Verbeke (1986, cited in Page, 1994) in her investigation into the pull factors of destinations. She showed in her analysis that water, canals and riverfronts needed to be highlighted separately from accessibility and were primary features of a destination. She also demonstrated in her model about tourism destination attractiveness, the importance of historical buildings alongside the water, and implicit within this is a vision of the past with very different types of structures for sometimes forgotten purposes. Such experiences provide an opportunity to consider the past and to step back from the everyday here and now. These now recognisably slow aspects of experience that we are now able to appreciate emerged in earlier historical accounts of spending time being on or near water. As Elizabeth Jane Howard (2002: 205) recalls, she:

> Grew to love canals and narrow boats: their secretive beauty, the way they slipped through large industrial towns and into country at a speed so leisurely – less than four miles per hour – that you could notice where you were . . .

Howard is endorsing Andrew's (2006: 145) view that a slower and more leisurely pace offers an opportunity 'to be aware of what is happening and what we're feeling' and that the opportunity to occupy time more attentively with slowness is mindful rather than mindless and demonstrates more acutely every day meaning and value (Parkins, 2004). Neal (1921) speaks of the contentment gained whilst cruising, where on the canals he has fresh air, sunshine, exercise, change of scenery, peace of mind and good food and an absence of news. He believes that being separated from the everyday events of the world helps him realise that it all carries on without him and that catching up after an absence is easily done. He also says:

> The word canal to the uninitiated conjures up a picture of an evil-smelling dirty stretch of water hedged in by the slums of any town, the only too obvious repository of the town's surplus bottles, tins and canine and feline quadrupeds. Poor souls! They have never seen that wonderful two miles of fairyland on the Shropshire Union canal from Llangollen to the Horseshoe falls, where every pebble in the canal bed shines through the pellucid water, where moss and ferns and flowers crowd on either bank, and the silver birch entwine their branches overhead in one long lovely avenue. (Neal, 1921: 10)

Parkins (2004: 365) writes of the very experience Neal describes, saying how slowness impacts on human behaviour and believes that it 'heightens aesthetic and sensory experience' and that spaces for slowness are to be allowed (Parkins, 2004: 367) in acknowledgment of the need for connection with places (Crouch, 2000). This view is substantiated again, this time by Haywood (2008: 65), who says 'the canals free us up in this way; they release us from the constraints that life imposes upon us'.

This recognition of the slowing down and connecting with local communities that takes place on canals links strongly with the sense of unity with nature, personal growth and rediscovered sense of self that is described as being integral to the slow tourism experience (Dickinson & Lumsdon, 2010).

Conclusion

Although travelling on and near canals is now better documented than ever before, there is a greater 'dislocation and disjuncture' in society (Hannam *et al.*, 2006: 160). This is because of the overwhelming volume of information that is now available. Added to this, much writing about transport history is too often examined in a way that disconnects it from the wider cultural milieu (Freeman, 1999), remaining in a world for specialists and enthusiasts. The very positive association that Sawday (2008) equates between slowness, happiness and wisdom, this author sees within research into canals and their users. Being slow in its broadest sense is integral to the whole experience of being on and near canals and makes explicit the sometimes unconscious desire of humans to take rest and relaxation. Slow movements create memorable arrival and departure points and strong sensory experiences blending with the people and the place (Dickinson & Lumsdon, 2010) are integral to every account of travel by narrow boat on canals.

The fast-moving pace of life, exacerbated by technological developments, has prompted a reaction: one that has recognised the dehumanising aspects and the resource implications. Whilst slow tourism may be the antithesis of mass travel, the discussion here has also shown how canal tourism and slow travel and tourism are synonymous. This has been done by addressing the canal context, including change of role over time, and by highlighting their connection with slowness, drawing upon authors' perspectives. It is hoped that this discursive approach challenges the supposedly required paradigm shift in thinking about tourism (Dickinson & Lumsdon, 2010) and purports a wider recognition and appreciation of our relationship with canals. Arguably, much of the leisure experience that takes place on or near canals epitomises slow travel and tourism.

References

Ackroyd, P. (2004) *Albion: The Origins of the English Imagination.* London: Vintage.

Andrews, C. (2006) *Slow is Beautiful: New Visions of Community, Leisure and Joie de Vivre.* Gabriola, BC: New Society Publishers.

Atterbury, P. (1994) *Exploring Britain's Canals.* London: Harper Collins.

Black, J. and MacRaild, D.M. (2000) *Studying History* (2nd edn). London: Macmillan.

Canal Cuttings (2011) Canal speed limit UK – Online document: http://www.canalcuttings.co.uk/boatownership/canal-speed-limit-uk.html

Carnegie, E. and McCabe, S. (2008) Re-enactment events and tourism: Meaning, authenticity and identity. *Current Issues in Tourism* 11 (5), 349–368.

Conway, D. and Timms, B. (2010) Re-branding alternative tourism in the Caribbean: The case for 'slow tourism'. *Tourism and Hospitality Research* 10 (4), 324–344.

Crouch, D. (2000) Places around us: Embodied lay geographies in leisure and tourism. *Leisure Studies* 19 (2), 63–76.

De Botton, A. (2003) *The Art of Travel.* London: Penguin Books.

Dennis, K. and Urry, J. (2009) *After the Car.* Cambridge: Polity Press.

Dickens, C. (2004) *American Notes.* London: Penguin Classics.

Dickinson, J. and Lumsdon, L. (2010) *Slow Travel and Tourism.* London: Earthscan.

Erfurt-Cooper, P. (2009) European waterways as a source of leisure and recreation. In B. Prideaux and M. Cooper (eds) *River Tourism* (pp. 95–116). Wallingford: CABI.

Fagence, M. (1998) Coping with tourists. *Annals of Tourism Research* 25 (1), 251–258.

Freeman, M. (1999) The railway as cultural metaphor: 'What kind of railway history?' revisited. *The Journal of Transport History* 20 (2), 160–166.

Graham, B., Ashworth, G.J. and Tunbridge, J.E. (2000) *A Geography of Heritage: Power, Culture and Economy.* London: Arnold.

Greenbie, B. (1981) *Spaces Dimensions of the Human Landscape*. New Haven, CT; London: Yale University Press.

Hadfield, C. (1986) *World Canals: Inland Navigation Past and Present*. Newton Abbot: David and Charles.

Hadfield, C. (1994) Foreword. In J. Boughey (ed.) *Hadfield's British Canals: The Inland Waterways of Britain and Ireland* (8th edn) (pp. i–v). Stroud: Alan Sutton Publishing.

Halsall, D. (2001) Railway heritage and the tourist gaze: Stoomtram Hoorn-Medemblik. *Journal of Transport Geography* 9 (2), 151–160.

Hannam, K., Sheller, M. and Urry, J. (2006) Mobilities, immobilities and moorings. *Mobilities* 1 (1), 1–22.

Harvie, C. (2002) Engineers holiday: LTC Rolt industrial heritage and tourism. In H. Berghoff, B. Korte, R. Schneider and C. Harvie (eds) *The Making of Modern Tourism: The Cultural History of the British Experience 1600–2000* (pp. 203–223). Basingstoke, Hampshire: Palgrave.

Haywood, S. (2008) *Narrowboat Dreams: A Journey North by England's Waterways*. Chichester, West Sussex: Summersdale.

Heidbrink, I. (2010) European canals. In J. Zumerchik and S. Danvers (eds) *Seas and Waterways of the World: An Encyclopedia of History, Uses and Issues* (pp. 118–121). Santa Barbara, CA: ABC-CLIO, Limited Liability Company (Greenwood).

Hoskins, W.G. (1981) *The Making of the English Landscape*. London: Penguin.

Howard, E.J. (2002) *Slipstream: A Memoir*. London: Macmillan.

Inland Waterways Association (2011) Bulletins – Online documents: http://www. waterways.org.uk/.

Jerome, J.K. (2007) *Three Men in a Boat: Not to Mention the Dog!* London: Penguin Popular Classics (original work published 1889).

Macbeth, J. (2000) Utopian tourists. Cruising is not just about sailing. *Current Issues in Tourism* 3 (1), 20–34.

Matless, D. (1994) Doing the English village, 1945–1990: An essay in imaginative geography. In P. Cloke, M. Doel, D. Matless, M. Phillips and N. Thrift (eds) *Writing The Rural: Five Cultural Geographies* (pp. 7–89). London: Paul Chapman Publishing.

Neal, A.E. (1921) *Canals, Cruises and Contentment*. London: Heath Cranton.

Owen, T.M. (1991) *The Customs and Traditions in Wales*. Cardiff, South Glamorgan: University of Wales Press Limitations.

Page, S. (1994) *Transport for Tourism*. London: Routledge.

Palmer, J. (2002) Smoke and mirrors: Is that the way it is? Themes in political marketing. *Media Culture Society* 24 (3), 345–363.

Parkins, W. (2004) Out of time: Fast subjects and slow living. *Time and Society* 13 (2–3), 363–382.

Reisinger, Y. and Turner, L.W. (2002) Cultural differences between Asian tourist markets and Australian hosts: Part 2. *Journal of Travel Research* 40 (4), 385–395.

Sawday, A. (2008) Go slow England. *The Guardian Travel Supplement*, 29 March.

Sawday, A. (2011) Sawday's ethical collection – Online document: http://www.sawdays. co.uk/about_us/ethical_collection/

Sawday, A. with McKenzie, G. (2008) *Go Slow England*. Bristol: Alastair Sawday Publishing Co. Ltd.

Squires, R. (1979) *Canals Revived: The Story of the Waterways Restoration Movement*. Bradford-on-Avon, Wiltshire: Moonraker Press.

Towpath Talk (2010) *Towpath Talk* 61 (4). http://www.towpathtalk.co.uk/

Trapp-Fallon, J.M. (2007) Reflections on canal enthusiasts as leisure volunteers. In F. Jordan, L. Kilgour and N. Morgan (eds) *Academic Renewal: Innovation in Leisure and Tourism Theories and Methods* (pp. 65–79). Eastbourne: University of Brighton.

Tresidder, R. (2000) The search for the sustainable. In M. Robinson, J. Swarbrooke, N. Evans, P. Long and R. Sharpley (eds) *Reflections on International Tourism: Environmental*

Management and Pathways to Sustainable Tourism (pp. 303–311). Sunderland: Business and Education Publishers.

UK Good Boating Behaviour code. Accessed 14 November 2011. http://www.apco.org.uk/images/the-Boaters-Toolkit.pdf

Vine, P.A.L. (1983) *Pleasure Boating in the Victorian Era.* Chichester, Sussex: Phillimore and Co.

Visit Britain (2011) Visit Britain. http://wwwvisitbritain.org

Part 4
Slow Tourism Places

12 Travellin' Around *On Yukon Time* in Canada's North

Suzanne de la Barre

The world's remote regions are spaces that are tied to a different conceptual geography and are perceived to exist in different time dimensions than 'central' areas (Ardener, 1989). They are spaces that remove us from the known, ordered, civilised and legible text, the modern, predictable, 'false' and fast world, and immerse us in the unknown, unreadable wild text, the pre-modern, unpredictable, 'authentic' and slow world. In remote regions, fast time is contrasted with an alternative slow time, which is often tied to the socially constructed pace of wilderness. The idea of wilderness time is immersed in myths about nature that emerged with the Romantic movement at the end of the 18th century, and was reinforced by the rapid industrialisation of Europe and many of its colonies, including Canada (Cronon, 1996; Hall, 2002; Jasen, 1995; Towner, 1996).

Remoteness is a powerful idea increasingly promoted for the purposes of tourism (Brown & Hall, 2000; Müller & Jansson, 2007; Wildlands League & Ontario Nature, 2005). Butler (2002: 7) highlights the allure of remote places and suggests that 'the more remote the location, the more valued it is as a collector's item'. The appeal of remote areas for tourists is significantly related to the images and myths associated with them. These include notions of discovering a still existing frontier, and the availability of unique and rare attributes that few others experience, such as meeting Aboriginal people, seeing exotic wildlife and being immersed in vast amounts of pristine wilderness.

A number of scholars have sought to better understand the desire for wilderness experiences through tourism. Wilson's (1991: 25) history of recreation and tourism in wilderness areas describes how nature was seen as a place one went to find 'some kind of contact with the origins of life' once urbanisation took hold. Jasen's (1995: 3) history of tourism in Ontario suggests that as 'fears about the effects of urban life mounted, [tourists] flocked from the enervating city to the exhilarating wilderness, hoping to cast themselves under the care of Mother Nature and to rediscover the power of the primitive within themselves'.

Dickinson and Lumsdon (2010) define slow travel as having two overarching dimensions: experiential and environmental. The experiential features include an appreciation for quality time, and for meaningful and locally engaged experiences that incorporate the actual physicality of slowing down. The environmental features focus on a commitment to travel that reduces carbon emissions; for instance, travel that replaces a reliance on air and car transport with travel by train and bus, or travel that is self propelled (e.g. cycling and walking). In addition, the environmental features of slow travel encourage travel that is closer to home, or that extends the time spent at a destination.

Place and pace characteristics associated with remote areas make them ideal slow travel destinations in terms of the experiential elements defined by Dickinson and Lumsdon (2010). However, remote destinations meet few of the identified environmental objectives. It is difficult for remote destinations to decrease carbon emission travel because of the distances required to reach these destinations, and the need for carbon emission travel upon arrival. Canada's Yukon offers an apposite example of a tourism destination that provides the experiential dimensions of slow travel. The Yukon's place-myths are associated not just with the slowed-down pace that is intrinsically associated with and experienced in peripheral settings; the territory's *timescape* also emphasises the idea that travel is an activity that has time transcendent qualities (Curtis & Pajaczkowska, 1994).

This chapter examines the relationship between remote regions and slow tourism by investigating how the experiential values associated with the slow movement are expressed by Yukon residents, as well as how the territory is marketed as a slow travel destination through its connection to those values. Through these examples, a number of issues will be examined that illustrate the Yukon tourism development context, and also highlight emerging issues and challenges faced by remote tourism destinations more generally. A qualitative interpretive approach is used to analyse Yukon resident place identity narratives and the way they are employed for destination marketing purposes. Narratives provide a compelling way to examine travel in relation to time (Curtis & Pajaczkowska, 1994; Germann Molz, 2009; Germann Molz, 2010).

Yukon resident place identity narratives and Yukon destination promotion campaigns are assessed for how they shed light on the associations made between place and pace (Shaw, 2001), and provide unique insights into slow travel and tourism, especially as it is encountered in remote areas. Textual data were collected from a variety of sources, including Yukon newspapers, northern magazines, as well as Yukon-made or made-about-Yukon documentary films. Textual data used to analyse how the territory is marketed for tourism include promotional material developed by the Yukon First Nations Tourism Association (YFNTA), and by the Yukon Government's Department of Culture and Tourism, specifically the 'On Yukon Time' campaign which was widely used from 2000 to 2005. The latter still has programme components that linger into the present (e.g. the annually updated *Art Adventures on Yukon Time* booklet).

The Yukon and its Tourism Development Context

Cushioned in the north western corner of Canada, the Yukon is 483,450 square kilometres and covers 4.8% of the country's total land mass (Figure 12.1). Alongside its vast space, the territory's remote and peripheral nature is characterised by many features, including: (1) its harsh climate (average minimum for Whitehorse in January is −22 degrees Celsius and −30.9 for Dawson City (Environment Canada, n.d.)) and northern light (midnight sun in summer, darkness in winter); (2) its low population density of 34,985 people, with almost 27,000 of these living in and around the Yukon's capital city, Whitehorse (Yukon Government (YG), 2010); (3) the presence of Indigenous people, referred to as First Nations in the Yukon; (4) its distance

Figure 12.1 Map of the Ukon Territory
Source: Reproduced with the permission of Natural Resources Canada 2011, courtesy of the Atlas of Canada.

to southern Canada and its amenities; and finally, (5) its limited, if ever-increasing, physical infrastructure.

In addition to these features, tourism development in remote destinations faces a number of challenges, including: (1) the relative absence of human and social capital (Hall & Boyd, 2005); (2) the low levels of accommodation; (3) seasonality issues; (4) economic factors; (5) training opportunities and the high number of minimum wage service sector employment; (6) the lack of industrial or manufacturing sectors; (7) insufficient markets for conventional development; (8) and weak political influence relative to core areas (Butler, 2002). A final and fundamental tourism development challenge faced by remote destinations is their limited accessibility. Accessibility issues in remote destinations are commonly defined by the lack of transportation options to and within them, making it difficult and/or expensive to get there (Müller & Jansson, 2007). Noting that distance dimensions have been the subject of many studies that model core-periphery relationships, Krakover (2002) instead highlights time dimensions in relation to the tourism development challenges remote regions face. He suggests that time and money are generally considered to be scarce resources, and that ordinary travellers seek to maximise time available at the destination, and minimise time en route to the destination.

Transportation issues are especially important for contextualising tourism development in the territory, and to understand also its positioning vis-à-vis slow travel. Territory-wide transport infrastructure has undergone significant changes over the last decade. In addition to 4700 kilometres of all weather, year-round roads which provide access to all Yukon communities except for one (Old Crow), both Air Canada and 'The Yukon's Own Airline', Air North, offer year-round service to major gateway cities (Edmonton, Calgary and Vancouver). Regional air service has expanded as well: Air North offers year-round flights to Dawson City and Old Crow, as well as flights to Fairbanks (Alaska) during the summer months. The newly expanded Whitehorse International Airport (completed in 2010) has, since 2001, offered direct flights from Germany on Condor Flugdienst during the summer months. Notwithstanding these developments, transportation issues present significant constraints for tourism development, especially the lack of options encountered once at the destination (e.g. there is no bus service between communities).

Narratives and Sense of P(l)ace

Slow travel values are embedded in popular Yukon place identity narratives. These narratives express renewed interest in the ideology of the sublime, rely on Romantic-like relationships to wilderness and nature, are affiliated with post-material values and share a critique of the cult of speed. The popular

What's Up Yukon weekly entertainment magazine regularly publishes articles that illustrate the kind of life pace embraced by Yukoners. In an article titled *Life in the Slow Lane is Canada's Best Commute* (Keating, 2007), a Dawson City resident explains how her daily commute to work from the cabin where she lives involves hooking up a dog or two and skijoring across the Yukon River into town. While most Yukon residents cannot claim a daily commute as exotic as the one described in this story, it aptly portrays the attachment many Yukoners have to an alternative life pace than the one thought to exist 'down south'.

Challenging Western belief systems and adopting 'alternatives' is a central theme also at the heart of these narratives. Transcending false ways of being is perceived as possible by disputing Western time and its related constraints:

> I quickly learned that the culture I grew up in is structured by 12-hour day-and-night cycles. Just think about it: In the Christian tradition, the Lord gave day to the living and night to the dead. The English language is peppered with idioms such as 'dark as night' and 'clear as day'. But those simply don't apply North of Sixty. (Sinclair, 2007: 76)

Newcomers to the Yukon express a variety of adaptations related to adjusted perceptions and experiences of time. For one new Yukoner, trade-offs connected with a move to the territory implicate a release from the clutter of modernity, and result in being forced to 'slow down half a notch' (in Black, 2005). Some relocated southerners confess to a period of initial discomfort:

> A frequent realization is that I've traded in tall buildings for stoic mountains: the bustle of the metropolis for the calm of wide open spaces. Ironically I felt a sense of panic over the fact that I needed to slow down upon my arrival. But as the North continues to grow on me, I find solace in the concept that it is more than welcome that you stop and breathe the mountain air. (McCarthy, 2007)

In the made-for-tourism promotion film that depicts what it is like to live in the Yukon, 'As the Crow Flies' (Chapman, 1998), one resident describes how time not only slows down in the territory, it moves backwards and recalls a childhood past: 'When I came here I totally fell in love with it. It was like going back to when I was a little girl.' Such narrative expressions insinuate that being in the Yukon can trigger a return to childhood: *the* quintessential original and authentic state. Childhood is a time also when, as proposed by Curtis and Pajaczkowska (1994), temporal and spatial dimensions are (still) experienced as integrated. 'Yukon time' can thus be understood as an alternative time that is synonymous with breaking down clock time, forward time.

Strang (1996: 64) offers a similar perspective on the way time is perceived in Australia's remote regions. She explores the tourist motivation to 'go bush' in Queensland, and explains how visitors perceive the region as a place that is 'apart from normal life, unspoiled, and uncorrupted by modernity'. Echoing the sentiment captured in Chapman's (1998) promotional film, Strang proposes that remote landscapes release people from authority and grown-up responsibility.

Alternative notions of time resonate differently with Yukon's First Nations people. Yukon historian Catharine McClellan (1987) explains how Aboriginal relationships to time, like past relationships to the land, were disrupted by newcomers and their ways of measuring and being in the world. Griffiths (1999: 341) describes how 'wild land and wild time' were discovered, charted and logged with the aid of theodolites, chronometers and telescopes, and points to how these were 'objects of finding in an unfoundland, inventions designed to find and log an unfoundtime'. In this way, newcomers brought relationships to the land which contributed to nation-building ambitions and rationales; purposes that relied also on the socially constructed idea of wilderness as empty landscape. Cronon (1996: 15) is not alone to advance the perspective that 'the myth of the wilderness as "virgin", uninhabited land had always been especially cruel when seen from the perspective of the Indians who had once called that land home' (for Canadian views see Coates & Morrison, 2005; Shields, 1991).

Hit the Brakes! Persuading Visitors to Slow Down

The Yukon's remote wilderness features coupled with its 1898 Klondike Gold Rush and First Nations history contribute to its reputation as one of the world's last great frontiers. The creation of a discourse that frames and sells Yukon wilderness as a tourist attraction is made evident as early as 1954 when a travel guidebook promoting the Alaska Highway proclaimed that it was still 'a frontier and unsettled region passing through the heart of an untamed wilderness' (Schreiner, 1992: 86). Little more than a decade later, one of the first Yukon tourism promotional films depicted 'the exciting places and events' a visitor could experience at a time when the territory 'still existed between frontier and modernity' (Willis, 1967). Affectionately referred to as the 'land of the midnight sun', the territory was already portrayed as offering an alternative tourism experience to urban visitors:

> But for those weary of crowded sky-scrapper cities, traffic jams, and smog, for those longing for sport and fun in an unspoilt wilderness, the Yukon is still the place to get away from it all. (Willis, 1967)

Somewhere down there is your SOUL

♣ CANADA'S YUKON   Yukon

Figure 12.2 Somewhere down there is your soul
Source: © Dan Heringa Photography. Retrieved on 29 February 2008 from http://www.
northendgallery.ca/somewheredownthere.html

Today, the territory attracts a variety of adventure-related tourists, and holds a particular appeal for visitors who seek spiritual or personal invigoration and renewal. Not surprisingly, the association made between remote wilderness and spiritual fulfilment is exploited by Yukon tourism marketing campaigns. Campaigns are designed to attract visitors by invoking emotional experiences that escort the visitor to the sacred and divine. An applicable example is found in 'artvertisements' like the one presented in Figure 12.2. Images like this one aim to persuade visitors that remote wilderness is a sanctuary where one can find one's soul cradled inside awesome space. The unarticulated assumption is that visitors arrive having lost touch with the sacred and the divine; that their urbanised lives predispose them to forget or misplace the sacred, the divine and, by extension, their souls.

In a similar manner, postcards and T-shirts that adorn Whitehorse storefront windows remind visitors that they can get 'Lost in the Yukon', thereby exploiting the rich imagery that links the territory to those rare places left on the planet where it is still possible to completely lose track of time and space. The state of being lost is called upon also as a crucial prerequisite for being able to find one's *true* (authentic, original, real) self. Solnit (2002) suggests that this yearned-for-lost is one that resonates with a similar lost mused upon by the legendary environmental philosopher Henry David Thoreau (1854: 268):

Not till we are completely lost ... do we appreciate the vastness and strangeness of nature. Not till we are lost, in other words, not till we have lost the world, do we begin to find ourselves, and realize where we are and the infinite extent of our relations.

Thoreau's lost inspires spiritual achievement. As Solnit (2002: 15) offers: 'Lose the whole world, get lost in it, and find your soul.'

To move slowly through place, and to find one's self through pace, are elements that resonate with the territory's *On Yukon Time* destination promotion campaign:

> Welcome to Canada's True North – a place where time itself is a little different. For Yukoners, it's a matter of pace, a way of conducting affairs so that life is experienced more fully, lived as it were ... on Yukon time. For some folks this means slowing things down so they can enjoy the moment. For others, it means heading out for a stroll or a paddle at midnight because the light is just right; taking a different route to see what the road holds in store; or adding a day to their itinerary to enjoy a local celebration. (Yukon Government, 2004a)

The campaign was used from 2000–2005 and was creative and strategic in the way it called upon the experiential values of slow travel, as well as the way it sought to position these values as essential Yukon place characteristics.

On Yukon Time operated with icons identifying places or events of interest, and persuaded visitors to 'slow down, explore and experience life on Yukon time' (Yukon Government, 2004a: 1). Visitors were further encouraged to:

> Hit the Brakes! Slow down and enjoy the moment. Take time to meet locals, discover special places. ... Relax. Breathe in our clean, fresh air. Spend a little extra time nourishing your soul. (Yukon Government, 2004b: 12)

The campaign also aimed to shift attention from the Klondike Gold Rush – which had long held the exclusive rights to the territory's tourism promotion efforts – to different tourism attributes worthy of visitor attention, including Yukon First Nations history and culture, for instance by visiting a First Nation cultural centre:

> Showcasing the traditional lifestyles of the Little Salmon/Carmacks First Nation, this cultural centre has fascinating exhibits. Enter a moose skin home or a brush house and imagine how life might have been. (Yukon Government, 2004a: 31)

Incentives such as activity passports that could be stamped at select venues were designed to get visitors to travel to communities outside Whitehorse and to take an interest in First Nations people.

First Nations people were displaced by the Klondike Gold Rush (Coates & Morrison, 2005; Cruikshank, 1992), and it is more than timely that this destination marketing shift indirectly addressed the tragically ironic fact that Yukon tourism relied for a long time on celebrating an event that had devastating and enduring negative impacts on Yukon First Nations people.

A Double-Edged Slow

Romantic longings for clocklessness and timelessness that are linked to the wilderness and ancient peoples are tied also to colonial histories. As an extension of this colonial past, the use of First Nations people in the Yukon's destination marketing reinforces also the association made between the territory, a cynicism towards and suspicion about the value of acceleration and a prejudice against speed itself. Jordan (1995) explains how the flight from modernity relies on the idea that some people are discursively fixed in time (belong to pre-history), while others develop through time (belong to the modern or post-modern). The presence of First Nations people provides a reference to pre-history and emphasises the positively framed anti-modern characterisations the territory holds for visitors to the Yukon (e.g. anti-acceleration).

These characterisations rely on notions of the primitive 'other', and serve to locate Yukon First Nations people in a historical past that is not always aligned with how they, as a contemporary and self-governing people, envision embracing their traditional history and finding ways to integrate elements of the past into the present. Nonetheless, slowing down is a theme that is used also in the tourism marketing materials created by the First Nations-driven not-for-profit association, the Yukon First Nations Tourism Association (YFNTA) (2005: 1):

Go slowly, take your time. Ask questions, and sometimes just listen. Pay attention to everything – the wind, the detail on a carving or the ripple on a river or a lake. Our history and our culture are all around you.

The imperative to appeal to visitors' desire to have meaningful interactions with local residents is one that benefits First Nations people, specifically where awareness and support for their contemporary cultural and political aspirations are concerned. Issues that surround complex representational issues and self-determination have captured the attention of tourism scholars interested in the ways that tourism, as a complex, multi-dimensional and global force, can negatively and positively impact Aboriginal people (Bunten, 2008, 2010; Butler & Hinch, 2007; Notzke, 2006).

Concealing Tourism Development Challenges in the Virtues of Slow Travel

The limited physical infrastructure and unpredictable access to amenities, both common physical characteristics of remote space, suggest that travel in remote destinations can provide actual opportunities to experience a slowed down world. The *On Yukon Time* campaign signals a desire to associate the Yukon with an alternative time-space continuum, and promotes the uniqueness and worth of the territory as a tourism destination. However, the idea of Yukon place-specific time supports also the view that unpredictable service encounters or lack of amenity options are part of what make the Yukon unique, in a positive way. The campaign creates an opportunity for destination marketers to entice tourists to 'reinterpret' infrastructural or customer service failings as an experience that is a natural – and welcome – result of 'northern lifestyle'.

Such 'reinterpretations' can be advantageous to Yukon tourism service providers who face geographic challenges that affect their ability to compete with southern-based service standards: for instance, the unpredictable nature of services due to weather, road conditions and lack of, or low capacity, human resources. These reinterpretations are beneficial also when visitors are faced with not being able to use a cell phone or the internet because there simply is no available service. Yukon communities have had cell phone and internet access since 2007; however, service outside community boundaries remains unavailable.

Another example of creative interpretations is found in the remote/accessible paradox that defines tourism in peripheral locations. Examples of strategies that are used to sell the Yukon as an accessible destination, while at the same time maintain its value and integrity as a remote place have resulted in promotional storylines such as this one:

> The Yukon's 483,450 square kilometres (302,156 sq mi) of wilderness lie above the 60th Parallel, north of British Columbia, east of Alaska and west of the Northwest Territories. As remote as this vast area appears on a map, it is actually easy and affordable to reach. (Yukon Government, 2004b: 14)

Marketers contend that remote places must at least be perceived as accessible; otherwise it is very difficult to attract visitors to them. As a result, 'accessible remoteness' is an increasingly used marketing theme. Branding research conducted by the territory in 2005 identified that the Yukon brand should emphasise the territory's wilderness and 'purity', its differentiation from the rest of Canada and its accessibility. Concomitantly, it stipulated that the branding strategy should avoid trying to portray it as too 'sophisticated' (e.g. city experiences and culture), and avoid also references to the territory's inaccessible/remote characteristics (Cameron Strategy Inc., 2005: 7). Consequently, tourism marketing has to negotiate a paradox that comprises two main points. First, marketing aims to convince visitors that the territory is accessible; the territory is close by and affordable and provides comfortable modern amenities. Second, marketing must maintain the perception that the Yukon is a remote place because this is what differentiates it from modernity and the fast world 'down south'. It is indeed a fine line to navigate. One way to make the best of a difficult situation, intentionally or not, is to engage the experiential qualities and attributes associated with slow travel.

Conclusion

Remote tourism destinations such as Canada's Yukon attract visitors because of the unique experiences they offer. The Yukon Government's *On Yukon Time* campaign highlights how the territory's promotional strategies have benefitted from the association that is made between the Yukon and the experiential values of slow travel. The campaign creatively and strategically positioned values associated with the slow movement as essential Yukon place characteristics. It is an association that was facilitated by the features of remote areas, and the slowed down timescape associated with them. The experiential dimensions of slow travel offer opportunities that can be advantageous to remote destination marketers. The Yukon case demonstrates how creative strategies can be designed to attract visitors who seek out nature, specifically extreme nature found in remote wilderness areas. Importantly, slow travel campaigns can also be designed to help navigate the remote/accessible paradox central to the remote destinations marketing problematic.

An underdeveloped area of research relates to the connection between Indigenous people and slow tourism. A more in-depth examination of the way slow travel is marketed in destinations where there are Indigenous people would be beneficial to understand the way in which ancient peoples and cultures are perceived and represented. These investigations can assist also in the critical examination of the challenges Indigenous people are faced with in their quest to represent themselves as living cultures, especially when engaging with complex tourism processes. Future research in this area will contribute to providing innovative strategies that assist Indigenous people to approach tourism development in a way that will be advantageous to them for economic purposes, as well as for social and cultural ones.

Finally, future research will inevitably be required to examine the relationship between slow travel in remote regions and the challenges posed by climate change and the unique issues it presents to peripheral destinations. Arguably, as Dickinson *et al.* (2010) claim, slow travel is not applicable to every tourism context. However, it is increasingly difficult to exploit the experiential elements of slow travel and tourism without addressing the features that relate to the need to lower carbon emissions.

References

Ardener, E. (1989) Remote areas – Some theoretical considerations. In M. Chapman (ed.) *Edwin Ardener: The Voice of Prophecy and Other Essays* (pp. 211–223). Oxford: Basil Blackwell.

Black, A. (Director) (2005) *Beautiful and Deranged: The Song of the Yukon* [motion picture]. Canada: Top of the World Films.

Brown, F. and Hall, D. (eds) (2000) *Tourism in Peripheral Areas*. Clevedon: Channel View Publications.

Bunten, A.C. (2008) Sharing culture or selling out? Developing the commodified persona in the heritage industry. *American Ethnologist* 35 (3), 380–395.

Bunten, A.C. (2010) More like ourselves: Indigenous capitalism through tourism. *American Indian Quarterly*, 1 July – Online document: http://www.faqs.org/periodicals/201007/2095927421.html

Butler, R. and Hinch, T. (eds) (2007) *Tourism and Indigenous Peoples: Issues and Implications*. Oxford: Butterworth-Heinemann/Elsevier.

Butler, R.W. (2002) The development of tourism in frontier regions: Issues and approaches. In S. Krakoer and Y. Gradus (eds) *Tourism in Frontier Areas* (pp. 3–19). Lanham, MD: Lexington Books.

Cameron Strategy Inc. (2005) *Consultations Report*. Yukon: Yukon Tourism Brand Strategy Development, Department of Tourism and Culture, Yukon Government.

Chapman, R. (Director) (1998) *As the Crow Flies: A Yukon Portrait* [motion picture]. Yukon: Whitehorse (Producers); Tourism Yukon; Yukon Government; Brandworks International and Robert Toohey Locations Ltd.

Coates, K. and Morrison, W. (2005) *Land of the Midnight Sun: A History of the Yukon* (2nd edn). Montreal, QC: McGill-Queen's University Press.

Cronon, W. (1996) The trouble with wilderness; or getting back to the wrong nature. *Environmental History* 1 (1), 7–27.

Cruikshank, J. (1992) Images of society in Klondike gold rush narratives: Skookum Jim and the discovery of gold. *Ethnohistory* 39 (1), 20–41.

Curtis, B. and Pajaczkowska, C. (1994) 'Getting there': Travel, time and narrative. In G. Robertson, M. Mash, L. Tickner, J. Bird, B. Curtis and T. Putnam (eds) *Travellers' Tales: Narratives of Home and Displacement* (pp. 199–215). London, New York: Routledge.

Dickinson, J. and Lumsdon, L. (2010) *Slow Travel and Tourism*. London: Earthscan.

Dickinson, J., Robbins, D. and Lumsdon, L. (2010) Holiday travel discourses and climate change. *Journal of Transport Geography* 18, 482–489.

Environment Canada (n.d.) Source Canadian climate normals 1971–2000 – Online document: http://www.climate.weatheroffice.ec.gc.ca/climate_normals/index_e.html

Germann Molz, J. (2009) Representing pace in tourism mobilities: Staycations, Slow Travel and The Amazing Race. *Journal of Tourism and Cultural Change* 7 (4), 270–286.

Germann Molz, J. (2010) Performing global geographies: Time, space, place and pace in narratives of round-the-world travel. *Tourism Geographies* 12 (3), 329–348.

Griffiths, J. (1999) *A Sideways Look at Time*. New York: Penguin Putnam.

Hall, C.M. (2002) The changing cultural geography of the frontier: National parks and wilderness as frontier remnant. In S. Krakover and Y. Gradus (eds) *Tourism in Frontier Areas* (pp. 283–298). Cumnor Hill, Oxford: Lexington Books.

Hall, C.M. and Boyd, S. (2005) *Nature-Based Tourism in Peripheral Areas: Development or Disaster?* Clevedon: Channel View Publications.

Jasen, P. (1995) *Wild Things: Nature, Culture, and Tourism in Ontario 1790-1914*. Toronto: University of Toronto Press.

Jordan, G. (1995) Flight from modernity: Time, the other and the discourse of primitivism. *Time & Society* 4 (3), 281–303.

Keating, L. (2007) Life in the slow lane is Canada's best commute. *What's up Yukon, My Dawson with Lulu Keating*, 30 November, p. 17.

Krakover, S. (2002) Time dimensions and tourism development in peripheral areas. In S. Krakover and Y. Gradus (eds) *Tourism in Frontier Areas* (pp. 21–37). Lanham, MD: Lexington Books.

McCarthy, T. (2007) Story corner, melodic initiation to the great outdoors. *What's Up Yukon*, 13 July, p. 3.

McClellan, C. (1987) *Part of the Land, Part of the Water: A History of the Yukon Indians.* Vancouver, BC: Douglas & McIntyre.

Müller, D.K. and Jansson, B. (eds) (2007) *Tourism in Peripheries: Perspectives from the far North and South.* Wallingford: CAB International.

Notzke, C. (2006) *The Stranger, the Native and the Land: Perspectives on Indigenous Tourism.* Concord, ON: Captus Press Inc.

Schreiner, J. (1992) Highway to Alaska: Fiftieth anniversary of the road that irrevocably changed the Canadian Northwest. *Canadian Geographic* 112 (2), 80–88.

Shaw, J. (2001) 'Winning Territory': Changing place to change pace. In J. May and N. Thrift (eds) *Timespace: Geographies of Temporality* (pp. 120–132). London: Routledge.

Shields, R. (1991) Chapter 4: The true North strong and free. In R. Shields (ed.) *Places on the Margin: Alternative Geographies of Modernity* (pp. 162–205). London: Routledge.

Sinclair, J. (2007) The unbearable darkness of being. *Up Here*, November–December, pp. 73–81.

Solnit, R. (2002) *A Field Guide to Getting Lost.* New York: Viking.

Strang, V. (1996) Sustaining tourism in far North Queensland. In M.F. Price (ed.) *People and Tourism in Fragile Environments* (pp. 51–67). Chichester: John Wiley & Sons Ltd.

Thoreau, H.D. (1854) *Walden: Or, Life in the Woods.* Boston, MA: Ticknor and Fields.

Towner, J. (1996) The search for wilder places. In *An Historical Geography of Recreation and Tourism in the Western World 1540–1940* (pp. 139–166). Chichester: John Wiley & Sons Ltd.

Wildlands League and Ontario Nature (2005) *Remoteness Sells: A Report on Resource-Based Tourism in Northwestern Ontario* – Online document: http://www.wildlandsleague.org/attachments/tourismreport.pdf

Willis, B. (Director) (1967) *Yukon: Getaway Country* [motion picture]. Vancouver, BC: Canawest Films.

Wilson, A. (1991) *The Culture of Nature: North American Landscape from Disney to the Exxon Valdez.* Toronto: Between the Lines.

Yukon First Nations Tourism Association (2005) *Visitor Guide 2004–2005.* Whitehorse, Yukon: Yukon First Nations Tourism Association.

Yukon Government (2004a) *Places to go on Yukon Time.* Whitehorse, Yukon: Department of Tourism and Culture.

Yukon Government (2004b) *Yukon: Canada's True North: Vacation Planner.* Whitehorse, Yukon: Department of Tourism and Culture.

Yukon Government (2010) *June 2010 Demographic Statistics.* Whitehorse, Yukon: Yukon Bureau of Statistics.

13 'Fast Japan, Slow Japan': Shifting to Slow Tourism as a Rural Regeneration Tool in Japan

Meiko Murayama and Gavin Parker

Several rounds of technological innovation have acted to speed up life experienced in the 21st century, not only in Japan but in many countries worldwide. Such technologies of speed have led to a variety of outcomes in terms of relations with place, the environment and in terms of psychological states, which are too complex to expound here (e.g. Virilio, 1986; Harvey, 1989; Hubbard & Lilley, 2002). However, speed, travel possibilities and communications have all had a marked impact on tourism and, more recently, concerns over social and environmental impacts of tourism have resulted in many voices arguing for 'slower' and more deliberative forms of behaviour.[1]

Slow tourism is a label that encapsulates part of this concern to re-emphasise the quality of experience and the impacts of tourism. It is possible to argue that slow tourism in Japan has many centuries of history during which infrastructures for slow travel and tourism were built up. Early leisure practices and tourism before the 20th century were mainly for health (e.g. spa) and religious purposes (e.g. pilgrimages to shrines and temples) and these forms of tourism were comparable to early European tourism activity (cf. Graburn, 2004). There was also an early form of slow 'business tourism' wherein travel to the capital *Edo* (Tokyo) was obligatory for the feudal lords and their large retinues. These *Sankin kōtai* trips were made in order to show fealty to the *Shogunate* (see Guichard-Anguis & Moon, 2009) and were so resource intensive that they sustained significant economic activity.

The rapid economic growth and development of Japan in the second half of the 20th century augured newer behaviours and significant characteristics of modern Japanese tourism were shaped at that time. It could be argued that Japan 'accelerated' up to and beyond the West in the period 1955–1990 in economic terms (Inglehart & Baker, 2000; Upham, 1987). In this same period most Japanese were not able or willing to take long holidays (compared to some Western populations). Even now, on average, only eight to nine days are taken in paid holiday, with employees using less than 50% of their entitlement (Ministry of Health, Labour and Welfare, 2010). The dominant work ethic had helped create a culture in which 'taking time off' was viewed as an extravagance and which could lead to dismissal (Boling, 1998). Instead, the 'cramming' of leisure experiences into short periods of holiday time became the dominant travel culture, where ostensibly such practices were designed to ensure 'value' and not to 'waste' scarce leisure time (Ravenscroft & Parker, 1999). These factors were contributors to the development of modes of 'fast' tourism, despite the longer history and infrastructures developed for slower tourism in the previous centuries.

Slow tourism and the associated 'new tourism' have been filtering into Japanese consciousness as part of a wider shift away from the 'fast' tourism culture. The adoption of the slow tourism concept in Japan demonstrates the challenges and opportunities for slow tourism given that in rural areas a slower pace of life and the 'hard' and 'soft' infrastructures for 'slow' tourism tend to remain. Furthermore, there is a need for innovation and in-migration to help revitalise the rural economy and society. There is some clear potential for tourist authorities, policy-makers and the range of leisure and tourism operators in drawing on the idea of slow tourism as a unifying idea to develop more sustainable tourism.

Japan as the 'Fastest' Tourism Nation: 'Slowly' Moving Away From the 'Fast' Model?

Japan has become the 'fastest nation' in the world according to Levine (2006), with the dominant culture and often cited *Kaizen* mentality (i.e. continuous improvement, efficiency and convenience) shaping Japanese attitudes to speed and efficiency (Noguchi, 1994). Knight (2010) notes that one of the clearest manifestations of 'fast Japan' can be seen in terms of travel and tourism. For example the 1000 km roundtrip between Tokyo and Kyoto is possible in a day due to the *Shinkansen* (bullet train). Before the railways were introduced this same journey would take around two weeks each way and travellers would stop in traditional inn towns, each with its own facilities and unique features that were developed over time.

Japanese habits and attitudes towards travel practices have been changing and fragmenting (Takai-Tokunaga, 2007) with many tourists taking advantage of a highly differentiated domestic and international market in tourism. Until recently group tourism was the main form of tourism in Japan with 47.5% of market share in 1964, dropping to only 7.1% by 2003 (see Japan Tourism Bureau, 2004; Nihon Kanko Kyokai, 2005: 82; Quiroga, 1990;

Sato, 2008a). For many Japanese tourists this format was about cramming experiences into tight schedules and value was often equated with quantity rather than quality of experience (Yamamoto & Gill, 1999). Such 'fast' tourism has left significant damage to the natural environment, can impact adversely on community, often provides shallow and unfulfilling holidays and only rarely adequately benefits local economies. Fast tourism often means that the quality of the experience is likely to be shallow, with little chance to build any connection or sense of 'inhabitation' with the visited location (see Cresswell, 2004; Ingold, 2000).

The dramatic decline in this model can be explained by changing consumer demands, which have become more mature and diverse. Other factors driving change include reduced costs of travel and better and more easily accessed information. Sato (2008b) analysed this and came to the conclusion that there is a shift underway from an industry-led to a consumer-led market (i.e. from a supply-led to a demand-led market). As part of this the *Chakuchi-gata*, or 'destination oriented approach' to tourism seems to be the direction that the Japanese market is moving in (see Oie & Kanai, 2008). This also points towards an emerging trend in niche or special interest tourism (see Aiida 2011; JTBWallet, 2011; T-Gate, 2011).

By the 1980s, the generation who had worked hard to regenerate the Japanese economy began to reconsider their 'live, work, play' balance (Murayama, 2004). By the turn of the new millennium only 30% of Japanese were satisfied in their work, even though most were surrounded by an abundance of material goods (Oswald, 2002). Until 1990 there had been financial stability and good job security. Economic recovery was the prime objective overall and the Japanese were told by the state that prosperity would bring happiness and satisfaction in life. Traditional values, perpetuated through Shintoism and Buddhism, of spiritual health and a giving society, were to some extent offered up at the altar of economic and material advancement (Watanabe, 1995). The phenomenon of *Karoshi* 'death by over-work' and concerns over high suicide rates were voiced (Iga, 1986). By 1979 attitudes towards material versus mental health reached parity status with 41% of Japanese people stating that they valued mental prosperity rather than material prosperity (rated at 40%). Since then the balance has shifted further with a 60/30% balance being recorded in 2010 (Naikakufu, 2010). This 'post-materialist' trend has also supported a rise in concern over human impacts on the environment (Aoyagi-Usui *et al.*, 2003; Inglehart & Baker, 2000).

By the late 1980s the Japanese economy faced a severe recession and some Japanese started to reflect on their lifestyles in the 'speed society' (Heiko, 1989; Iga, 1986). This may be part of a generational shift: Japan has effectively moved through two generations where the lives of the adult working populations were dominated by the pressure to rebuild and strengthen Japan after World War II. The younger generation are tending to question the price paid to achieve economic success in personal, social and environmental terms. Some Japanese critics were alarmed at the reality of a society which seemed to have forgotten its traditional values for the sake of material wealth and money (see Iida, 1980; Nakano, 1992; Teruoka, 1989). Such a history and a shift in understanding of the 'contract' offered in the post-war period were shaped by the post-bubble economy prospect of job insecurity, increased unemployment and economic stagnation.

Changing attitudes amongst some segments of the population have also resulted in an increase in people returning to the countryside, labelled the 'U-turners', and others moving into the countryside from urban areas, the 'I' or 'J-turners' (National Institute of Population and Social Security Research, 2010). National surveys have found that such 'turners' are seeking a slower lifestyle and are placing value on daily life and everyday experiences (Hidaka, 2003; Kainan City Council & RILG, 2002). This situation has had an effect on the tourism market too. There has been a shift towards valuing deeper experiences and, for example, an interest in working holidays in fishing, agriculture or forestry has developed. A large number of urban tourists are coming to the countryside where they actively look to participate in rural activities, typically through farm-stays or weekend trips (Creighton, 1997).

From New Tourism to Slow Tourism in Japan

There has been a widespread interest in new forms of tourism since the 1990s. A range of terms have been coined and one umbrella label used is that of the 'new tourism' or 'alternative tourism' widely discussed in the Anglophone literature (e.g. Butler, 1990; Poon, 1993; Smith & Eadington, 1992; Wheeller, 2003). Green tourism as a label was first introduced by the Japanese Ministry of Agriculture, Forestry and Fisheries (MAFF) in 1992 as a means of revitalising rural Japan (Aoki, 2010). The first Ecotourism Association was established in 1996 in Okinawa and the Japan Ecotourism Society was founded in 1998 to promote more sustainable tourism forms. More recently the Ecotourism Promotion Enforcement Act (2008) was passed, led by the Ministry of the Environment, and green or ecotourism forms have become institutionalised and sponsored at the national level. The Ministry of Land, Infrastructure and Transport (MLIT) and The Japan Travel Agency (JTA) have been promoting 'New Tourism' since 2007, which includes: ecotourism, green tourism, health tourism, long-stay tourism, industrial heritage tourism and cultural tourism. The JTA has chosen various local tourism projects to support the growth of such 'new tourism' (Japan Travel Agency, 2010). Recently other terms such as 'food tourism' have also been recognised and are in some instances being subsumed or linked with slow tourism (Hall et al., 2003).

Slow tourism is a rather newer term and has not gained wide recognition in Japan as yet and there is a relative scarcity of slow tourism literature generally (although, see Peeters et al., 2006) and consequently the concept of slow tourism has no clear or agreed definition. A loose understanding of valuing slowness is fairly obviously associated with the term; stemming from the Italian slow food movement emerging in the late 1980s (see Petrini, 2003). Some consideration of slow tourism has been emerging in recent years in the Japanese literature. Naito (2006) and Uchida (2006) agree that 'slow tourism' is an oppositional construct aimed at 'unsustainable' lifestyles and tilted against forms of 'fast' tourism indicated above. Takeuchi (2005) emphasises slow tourism as a positive future model, as opposed to mass tourism. In Japan the terms 'slow life' and 'slow food' are more recognised and have been applied to promote a slower pace of life. The wonderfully labelled 'Sloth Club' was established in 1999 and the 'Slow Life Japan' organisation was set up in 2003 with 'Slow Food Japan' launched in 2004. These third sector initiatives have been established to introduce and permeate the idea of 'slowness' as a virtue (see Slow Food Japan, 2010; Slow Life Japan, 2010; The Sloth Club, 2010). Some municipalities, such as Kakegawa City also now use the term 'slow' as a key term to promote slower lifestyles (see, Kakegawa City Council, 2010; NPO Slow Life Kakegawa 2010).

In attempting to provide a working definition of slow tourism Sugiyama and Nobuoka (2007: 3) highlight it as the 'type of trip that enables self realisation through doing or being slow, enabling closer observation rather than simply sightseeing'. They set five conditions or characteristics for slow tourism:

(1) Being healthy and involving walking and the enjoyment of 'slow food'.
(2) Involving at least one overnight stay.
(3) Featuring opportunities for self-realisation.
(4) Limited use of cars or other motor transport.
(5) To be in some sense green or ecological.

Takeuchi (2005) characterises slow tourism as being unhurried, high quality with overnight stay(s) and with experience or activities that are 'locally produced and locally consumed'. An early example of slow tourism research in Japan was carried out by Chugoku Regional Transport Bureau (CRTB), which conceptualised the term as follows: 'slow tourism is a type of travel with overnight stay(s) taken by individuals or small groups spending time slowly to interact with local people or/and nature' (Chugoku Region Transport Bureau, 2009: 1). Sugiyama and Nobuoka (2007) conclude that green tourism, ecotourism and agri-tourism may be seen as part of the slow tourism construct, i.e. that the label may be viewed as an umbrella term. Takeuchi (2005) notes that the resources and spaces of agri-tourism and green tourism are intrinsically involved in slow tourism, as agriculture, forestry and fisheries often provide the locations, experiences and produce required to fulfil the requirements of a slow tourism experience. These definitions are wide and allow for a wide range of 'tourisms', yet together they give an indication of the types of features and experiences likely or possibly associated with this umbrella term. This highlights how slow tourism already acts as a middle-range label that stretches across existing forms of niche tourism and as part of the wider 'new tourism'. We draw together the idealised features or dimensions of slow tourism set against 'fast' tourism in Table 13.1.

Thus far 'slow' tourism is only slowly gaining ground in Japan. The JTA promotes and uses the term 'new tourism' instead and it appears that for many the label is not considered strong enough to supplant existing green, ecotourism, food tourism, or related terms which encompass some or all of the practices typically associated with 'slow' tourism. At the time of writing no national governmental ministries use 'slow' as a key policy term.

Table 13.1 Features of fast tourism and slow tourism

Dimensions	Fast tourism	Slow tourism
Relationship	Instant	Harmonious
Ownership	Outside capital	Locally owned, individual/ cooperative
Local people	Marginalised	Main actors
Scale	Often larger	Mainly smaller
Size	Accommodate larger groups	Mainly individuals, families or small friendship groups
Activity	Hoping/passive	Doing/active/being
Value	Efficiency, quantity	Valued experiences, quality
Pattern	Standardised	Flexible
Process	Tourism industry led	Bottom up/co-constructed
Pattern	Standardised	Flexibility
'Inhabitation'	Not possible	Possible
Options	Many places	Selective
Mentality	Taking	Absorbing, deep appreciation
Spirituality	Consumerist, materialistic	Fulfilment, purifying
Benefits	Few larger businesses	Wider community
Local economic leakage	More leakage	Minimised leakage
Length	Short stay	Longer stay
Sustainability	Unsustainable	Sustainable
Examples	Group package tours, day visits	Ecotourism, green tourism, agri-tourism, health tourism, cultural tourism, food tourism

Some local government authorities at regional, prefectural and municipal levels appear to be making use of this new term, but it also appears that they may not fully understand what 'slow' really implies, or what would need to be done to realise this form of tourism. Instead it has been used as a temporary promotional tool or applied to other aspects of life. Since late 2010 Kyoto City has started to promote the 'Slow life Kyoto' message to shift the car-oriented city towards a more pedestrian-oriented lifestyle (Kyoto City Council, 2011), but it has not adopted this for tourism. Although some other municipalities (local authorities) have started to use the term in policy and planning documents, it is often a slogan used without detailed strategies or implementation plans (see Kakegawa City Council, 2010; Myoko City Council, 2010). A notable exception is Gifu City which now has one of the longest histories of applying the 'slow' concept into its policies and it is now expanding this to 'slow music', 'slow education', 'slow food' and 'slow industry' (Gifu City Council, 2010). Evaluation of the tourist information provided on various websites and visitor leaflets based in and around Gifu shows that the 'slow tourism' tag is not overtly or widely used however, and the clarity of the message and power of the label of 'slow tourism' is therefore somewhat diminished (Gifu Convention and Visitor Bureau, 2010).

In terms of the private and third sectors the idea of the 'slow' life is often used to help sell products with some environmental credentials; there are groups who take this notion seriously and engage in various activities in the countryside to promote a slower way of life and use the slow life slogan to encourage visitors or even new residents ('I', 'J' or 'U' turners) into the countryside (see Slow Life Centre, 2011). Our reflection is that the term is competing with other more specific terms and constitutive practices in Japan. As such, many economic development and tourist authorities are unsure about whether to replace existing marketing labels with slow tourism, preferring to use the term in policy circles rather than as a marketing tool for tourists. The next section focuses on examples of various levels of implementation of the slow tourism idea and why it is being discussed seriously in Japan.

Slow Tourism as a Tool for Rural Regeneration in Japan

Economic revitalisation is one of the most serious and urgent priorities for rural regions and the role of the public sector in regenerating the rural economy is seen as essential. The economic and demographic divergence between urban and rural areas has been widening and many rural areas feel they are 'sinking' or in a downward economic and social spiral. The key challenges are depopulation (migration into cities primarily) and an ageing rural population. Therefore, the means to revive the rural economy and make it attractive both to 'I', 'J' and 'U-turners' and to visitors are viewed as critical. As part of this, tourism is regarded as one of the main hopes to reinvigorate rural areas. It has been national policy for two decades to encourage urban residents to visit the countryside. Thus, as part of this a drive towards expanding 'green tourism' has been strongly supported by various national government departments (Aoki, 2010). There is also an idea shared by many local economic development officials that by encouraging visits and holidays of certain types (i.e. new tourism forms) some tourists (as 'turners') will be tempted to live or work in those rural areas, so tourism becomes a means of attracting new residents. The concern for Japanese policy-makers is to shift tourism forms towards assisting in shaping a more sustainable rural economy and society overall.

Despite interest in new tourism generally slow tourism research is still in its infancy. There are only two major research studies that examine this form of tourism in Japan. These were conducted in Chugoku Region, (see Figure 13.1) in the western part of Japan. These evaluative studies were funded by the national government through the Ministry of Land, Infrastructure, Transport and Tourism (MLIT) and the Chugoku Regional Transport Bureau (CRTB) (Chugoku Region Transport Bureau, 2006, 2009). The region still retains its rural character and an extensive rural landscape. It is also suffering from a range of socio-economic problems, in common with many rural areas in Japan. The research studies evaluated how and whether 'slow tourism' could be an

Figure 13.1 Japan and the Chugoku region

effective tool to revitalise the economy from the bottom up and at the same time proposed practical suggestions for the implementation of slow tourism. The first study by the Chugoku Regional Research Centre (CRRC) picked 12 cities to implement slow tourism (see Chugoku Regional Research Centre, 2010). However, none of the 12 cities yet use 'slow tourism' explicitly on their own official websites for visitors and the term cannot be found on any visitor information websites. However, two prefectures in the Chugoku region have used the slow tourism term. Kagawa Prefecture used it only for a short time on its website to casually promote slow tourism. According to the local tourism association, slow tourism was never a recognised or formally adopted term there (Kagawa Tourism Association, 2010a, 2010b). On the other hand Yamaguchi Prefecture is one of the places which has been actively promoting slow tourism since 2007 and this case is discussed below.

The second research project conducted by Chugoku Regional Transport Bureau (2009) focused on the rural areas of Chugoku region specifically and evaluated the possibility of implementing and facilitating slow tourism. Emphasis was placed on the choices of experiences and activities offered to tourists, which would enable longer 'slower' stays. Among the 64 municipalities which participated in the research (from a total of 110) only 20 had employed the term 'slow tourism'. Some others were only just starting to discuss this approach and so the number of municipalities providing or using slow tourism as a key promotional tool by the end of 2009 was low. The review indicated that Yamaguchi was the only prefecture taking slow tourism forward as a main policy approach for tourism and economic development.

Yamaguchi is located on the periphery of *Honshu*, the main island of Japan. The rural areas of Yamaguchi lost 32% of its population from 1950–2005 (Yamaguchi Prefecture, 2007: 6) while increasing numbers of urban visitors have enjoyed activities and experiences in rural Yamaguchi. Over 1.56 million urban residents visited rural areas in the prefecture in 2003 (Yamaguchi Prefecture, 2007: 10). Based on past experience the prefecture now emphasises tourism activities in its rural areas and since 2007 they have viewed 'slow tourism' as a key mechanism to further promote activity and experience-based tourism. The prefecture uses slow tourism as an umbrella term to include various activities taking place in rural areas. The prefecture has placed slow tourism as the 'next step on' from various forms of ecotourism (Cater & Cater, 2007; Yamaguchi Prefecture, 2007), which they have actively promoted in the recent past. They understand that slow tourism can be used to bring together a range of activities, products and experiences to useful economic advantage. They encourage the sales of locally produced goods, as well as trying to encourage the 'turners' to come and live in the prefecture.

Thus Yamaguchi emphasises slow tourism to promote economic activities in rural destinations and to market various types of activities involving agriculture, fishing or forestry as important attractors (Yamaguchi Prefecture,

2007, 2010). The Yamaguchi slow tourism vision encompasses a holistic approach to revitalise and utilise rural resources. Visitors can stay at bed and breakfasts (B&Bs) with farmers, fishermen and forestry families, or eat at farmers' and fishermen's own restaurants, attend local farmers' markets and engage in various other activities in the rural areas (Yamaguchi Prefecture, 2010). In order to achieve this 'slow' vision rural destinations are encouraged to create attractive tourism resources and develop new tourism businesses. A need for investment in human resource development was identified as part of this programme (Yamaguchi Prefecture, 2007) and the prefecture also identified the establishment of support organisations as being necessary to help promote slow tourism and products to relevant markets. A dedicated section of the prefectural government guides this work and runs the website used to promote slow tourism.

The Yamaguchi slow tourism policy aims to satisfy both residents in rural areas, as well as visitors from urban areas. Rural residents are seen as important providers of various services and products and the approach represents a form of neo-endogenous economic development (Ray, 2006). The interaction and economic benefit is supposed to generate mutual value and appreciation. This micro-economic activity is also designed to help bring extra income to rural households (and to female householders in particular, see Yokohama & Sakurai, 2006). Facilities such as *Michinoeki* 'road side stations' (see Parker, 2010) and *Chokubaijo* 'farmers' markets', and other outlets provide for local rural products and experiences and typically feature direct involvement of local craftspeople, farmers and others. These have become key centres for urban visitors, where information is given and local products are sold or unique experiences are offered or signposted. Yamaguchi's slow tourism offer also encourages school trips for urban students to learn about the lives of rural people and their work in the countryside. Where possible they stay with families to gain more understanding of their lives. Slow food is also an essential element. The Japanese tend to be very interested in food and in the diversity of local food cultures. New types of restaurants and cafes have been emerging in Yamaguchi where locally sourced or produced foods are cooked by farmers, fishermen or forestry families. This type of practice has been encouraged by agricultural, forestry and fishing policies which promote locally produced and locally consumed produce. Indeed, the direction of this policy in Yamaguchi is supported by shifts in regional consumer preferences, as discussed at the outset of the chapter. For example 48% of urban residents preferred to stay in farmers' or fishermen's houses in 1998, but by 2005 this had risen to 69% (Yamaguchi Prefecture, 2010: 8).

Although Yamaguchi uses the 'slow tourism' label, similar efforts towards rural regeneration can be found all over Japan, as such it is possible to say that slow tourism is more widely promoted but is done so in other guises. While there are less developed or integrated tourism resources in rural areas there are valuable tourism assets that could be used better (Chugoku Region Transport

Bureau, 2006; Japan Association of Travel Agents, 2004) and there is plenty of scope to make inroads using slow tourism as a discursive resource and framework applied to create synergies between different producers and tourism actors. For example, this could include making use of various skills and experiences of retired people in rural areas. For example, on *Iki-no-shima* island, in the south west of the Chugoku region, a volunteer action group, 'Team Sakimori', has been set up to revitalise the island's economy which has had a heavy dependence on traditional tourism. Most of the active members are retired residents including 'I', 'J' and 'U' turner residents. They organise volunteers to clean up rubbish on the beaches, support local festivals and provide tour guide services to individual tourists (Team Sakimori, 2010). The role that such groups may play in the development of slow tourism clearly has potential and drawing on local knowledge, effort and networks by other strategic actors to form partnerships could be even more beneficial and is less well orchestrated in much of Japan (see Parker & Murayama, 2005).

Conclusion: Just Another 'Naming Game' or the Ultimate Direction?

The use of slow tourism as a label is still in its first steps in Japan although other marketing efforts and approaches are well developed using other 'new tourism' labels. Thus slow tourism is not always seen as being useful as a marketing tool when there are other labels already being deployed. Others are still preferred by national policy-makers and there is some potential for fragmentation of effort here, for example 'green tourism' is widely promoted by MAFF, 'ecotourism' by the Ministry of Environment, and 'new tourism' by MLIT and JTA. These forms or labels for tourism activity do however seem to indicate a departure from what we have referred to as 'fast' tourism.

These new labels and activities may be grouped and brought together, with other activities following the loose definitions provided above under the banner of 'slow tourism' to bring together a more sustainable tourism. A genuinely slow tourism will need persistent and widespread effort if such a shift is to help assist in the social and environmental readjustment of Japan into the 21st century, yet the real potential of the label lies in helping to harness and integrate different activities and resources that may help deliver more sustainable rural economies and societies. Yamaguchi Prefecture is one of the most proactive areas in Japan where the term 'slow tourism' is actually used. Though some effort has been made to promote locally produced and activity-based tourism services, more innovative, localised or organic approaches may be needed for sustainable slow tourism to spread (see Aiida, 2011; JTBWallet, 2011; T-Gate, 2011, Yamaguchi Prefecture, 2010). As part of this the role of the community in planning, distributing and providing services and managing all elements of 'slow tourism' is likely to be important and the role of the existing tourism industry also needs significant change as traditional tourism and 'fast' tourism are still strong in Japan. Thus, the accumulation of knowledge, experience and skills required to move towards slow tourism (and a 'slow' rural economy) by each community is a longer term process. This is a lesson that Japanese public officials are only now beginning to understand.

A deeper absorption of the slow tourism philosophy is contingent on a number of factors. In our view slow tourism will take more time and consistent effort and development of understanding from providers, promoters and consumers. Otherwise slow tourism will remain as just another label sitting alongside the rest of the niche or new tourism labels used by a few authorities seeking to freshen their marketing. If this happens it would be a great pity as the history of Japan and the residual culture that existed before 1945 is well suited to the slow tourism approach and much infrastructure and goodwill already exists. Moreover there is clearly a cultural association that is being reformed between rural areas and a search for a 'better' life on the part of a significant number of urban dwellers and the 'turners'. The synergy effect of drawing together the various resources and possibilities that Japan's tourism industry and local entrepreneurs possess presents a very powerful opportunity. It is clear however that more research to understand how slow tourism is being organised on the ground in Japan and what is needed to actualise and develop this in partnership with local actors is needed (Soshiroda, 2010). Indeed, international understandings of the conditions and behaviours of slow tourists and other actors necessary for developing slow tourism are also clearly required.

Note

(1) This chapter was at final draft stage when the East Japan Earthquake and subsequent tsunami struck in March 2011. It is worth reflecting that, despite the Japanese living in one of the most severe natural disaster zones in the world, much of the wisdom of living in harmony with nature appears to have been marginalised in the 'fast' era. It is to be hoped that the underpinning philosophy of ideas such as slow tourism can point toward a new direction; towards a re-harmonisation with nature. This chapter is dedicated to all those affected by this disaster.

References

Aiida (2011) *Tikitabi* – Online document: http://tikitabi.com/

Aoki, T. (2010) *Tenkan suru Green Tourism (Changing Green Tourism)*. Tokyo: Gakugeisha.

Aoyagi-Usui, M., Vinken, H. and Kuribayashi, A. (2003) Pro-environmental attitudes and behaviours: An international comparison. *Human Ecology Review* 10 (1), 23–31.

Butler, R. (1990) Alternative tourism: Pious hope or Trojan horse? *Journal of Travel Research* 28 (3), 40–45.

Boling, P. (1998) Family policy in Japan. *Journal of Social Policy* 27, 173–190.

Cater, C. and Cater, E. (2007) *Marine Tourism: Between the Devil and Deep Blue Sea*. Wallingford: CABI.

Chugoku Region Transport Bureau (CRTB) (2006) Kankyofuka ni hairyosita Setonaikai slow tourism soushutu-chousa (Research on Setonaikai slow tourism considering environmental impacts) – Online document: http://www.mlit.go.jp/kokudokeikaku/souhatu/h17seika/3setouchi/3setouchi.html

Chugoku Region Transport Bureau (CRTB) (2009) Chuugoku chihou no slow tourism no teian (Proposal of slow tourism for chugoku region) – Online document: http://wwwtb.mlit.go.jp/chugoku/kikaku/slowtourism.html

Chugoku Regional Research Centre (CRRC) (2010) Setouchi slow tourism – Online document: http://www.chugoku-navi.jp/slow/index.html

Creighton, M. (1997) Consuming rural Japan: The marketing of tradition and nostalgia in the Japanese travel industry. *Ethnology* 33 (3), 239–254.

Cresswell, T. (2004) *Place: An Introduction*. Oxford: Blackwell.

Gifu City Council (2010) Slow life city Gifu: Welcome to the world of slow life – Online document: http://www.city.gifu.lg.jp/c/40121751/40121751.html

Gifu Convention and Visitor Bureau (2010) Let's travel around Gifu – Online document: http://www.gifucvb.or.jp/index.php

Graburn, N. (2004) The Kyoto tax strike: Buddhism, Shinto and tourism in Japan. In E. Badone and S. Roseman (eds) *Intersecting Journeys* (pp. 125–139). Chicago: University of Illinois Press.

Guichard-Anguis, S. and Moon, O. (eds) (2009) *Japanese Tourism and Travel Culture*. London: Routledge.

Hall, C.M., Sharples, E., Mitchell, R., Macionis, N. and Cambourne, B. (eds) (2003) *Food Tourism Around the World*. Oxford: Butterworth-Heinemann.

Harvey, D. (1989) *The Condition of Postmodernity*. Oxford: Blackwell.

Heiko, L. (1989) Some relationships between Japanese culture and just-in-time. *The Academy of Management Executives* 3 (4), 319–321.

Hidaka, N. (2003) Wakamono no chihoukaiki (Young people returning). *Best Value* 4 (12), 28–31.

Hubbard, P. and Lilley, K. (2002) Pacemaking the modern city: The urban politics of speed and slowness. *Environment and Planning 'D': Society and Space* 22 (2), 273–294.

Iga, M. (1986) *The Thorn in the Chrysanthemum*. Los Angeles: University of California Press.

Iida, T. (1980) *'Yutakasa' towa Nanika? (What is 'Prosperity?')*. Tokyo: Kodansha.

Inglehart, R. and Baker, W. (2000) Modernization, cultural change, and the persistence of traditional values. *American Sociological Review* 65 (1), 19–51.

Ingold, T. (2000) *The Perception of the Environment*. London: Routledge.

Japan Association of Travel Agents (JATA) (2004) Saranaru Kokunai Ryokou ni Mukete (For increased domestic tours) (Report) – Online document: http://www.jata-net.or.jp/membership/info-japan/research/index.html

Japan Travel Agency (JTA) (2010) New tourism soushutsu ryutsu sokushin jigyou (New tourism creation, distribution and support project) – Online document: http://www.mlit.go.jp/kankocho/shisaku/sangyou/new_tourism.html

Japan Tourism Bureau (JTB) (2004) *Tourism Marketing Institute News Release #1*, 11 November.

JTBWallet (2011) JTB Kankoujouhou navi (Tourism information navigation) – Online document: http://jtbwallet.jp/shop/default.aspx

Kagawa Tourism Association (2010a) Enjoy Kagawa – Online document: http://www.21kagawa.com/udon-t/index.html#

Kagawa Tourism Association (2010b) My Kagawa trip – Online document: http://www.my-kagawa.jp/

Kakegawa City Council (2010) Kakegawa City Top – Online document: http://lgportal.city.kakegawa.shizuoka.jp/

Kainan City Council and Research Institute for Local Government (RILG) (2002) Wakaisedai ni miryokuaru chiikisouzou ni kansuru kenkyu (Research on creating attractive localities for young generations) (Report). Tokyo: Nippon Zaidan.

Knight, J. (2010) The ready-to-view wild monkey – The convenience principle in wildlife tourism. *Annals of Tourism Research* 37 (3), 744–762.

Kyoto City Council (2011) Aruku-machi Kyoto Suishinsitsu (Kyoto as walking city promoting section) – Online document: http://www.city.kyoto.lg.jp/tokei/soshiki/9-5-0-0-0.html

Levine, R. (2006) *A Geography of Time: The Temporal Misadventures of a Social Psychologist, or How Every Culture Keeps Time just a Little bit Differently.* Oxford: Oneworld Publications.

Ministry of Health, Labour and Welfare (MHLW) (2006) Roudou-toukei-youran (Labour statistics) – Online document: http://www.mhlw.go.jp/toukei/youran/index-roudou.html

Murayama, M. (2004) *Understanding Urban Tourism.* Tokyo: Bunshindo.

Myoko City Council (2010) Bio-region – Online document: http://www.city.myoko.niigata.jp/

Naikakufu (Cabinet Office, Government of Japan) (2010) Kokuminseikatsu ni kansuru yoron chosa (Survey on lives of the nation) – Online document: http://www8.cao.go.jp/survey/h22/h22-life/z37.html

Naito, N. (2006) Slow tourism no genjo to tenbo (Current and future of slow tourism). In Ryoko-sakka no Kai (ed.) *Slow Tourism no Tenbo (Perspectives on Slow Tourism)* (pp. 12–23). Tokyo: Gendai Ryoko Kenkyujo.

Nakano, T. (1992) *Seihin no shisou (Thoughts On a Simple Non-egoistic Life).* Tokyo: Soushisha.

National Institute of Population and Social Security Research (NIPSSR) (2010) Population movement survey 2006 – Online document: http://www.ipss.go.jp/ps-idou/j/migration/m06/point.pdf

Nihon Kanko Kyokai (2005) *Kanko-no-jittai-to-shiko (Facts and Orientation of Tourism)* 23. Tokyo: Nihon Kanko Kyokai.

Noguchi, P. (1994) Savour slowly. Ekiben: The fast food of high-speed Japan. *Ethnology* 33 (4), 317–330.

NPO Slow Life Kakegawa (2010) Slow life info – Online document: http://slowlife-k.sakura.ne.jp/index.php

Oswald, A. (2002) Are you happy at work? Job satisfaction and work-life balance in the US and Europe. Paper presented at the WBS event, New York, 5 November – Online document: http://www2.warwick.ac.uk/fac/soc/economics/staff/academic/oswald/finalnywarwickwbseventpapernov2002.pdf

Oie, T. and Kanai, M. (2008) *Chakuchigata Kankou (Tourism from Destinations).* Tokyo: Gakugeisha.

Parker, G. (2010) Michi-no-eki: An opportunity for the rural economy? *Town and Country Planning* July–August, 342–346.

Parker, G. and Murayama, M. (2005) Doing the groundwork? Transferring a UK environmental planning approach to Japan. *International Planning Studies* 10 (2), 105–128.

Peeters, P., Gössling, S. and Becken, S. (2006) Innovation towards tourism sustainability: Climate change and aviation. *International Journal of Innovation and Sustainable Development* 1 (3), 184–200.

Petrini, C. (2003) *Slow Food: The Case for Taste.* New York: Columbia University Press.

Poon, A. (1993) *Tourism, Technology and Competitive Strategies.* Wallingford: CABI.

Quiroga, I. (1990) Characteristics of package tours in Europe. *Annals of Tourism Research* 17 (2), 185–207.

Ravenscroft, N. and Parker, G. (1999) Regulating time, regulating society: Control and colonisation in the sphere of public leisure provision. *Society and Leisure* 22 (2), 381–402.

Ray, C. (2006) Neo-endogenous rural development in the EU. In P. Cloke, T. Marsden and P. Mooney (eds) *Handbook of Rural Studies* (pp. 278–298). London: Sage.

Sato, K. (2008a) Mass Tourism-no-ikizumaari (Stagnation of mass tourism). In M. Inoue (ed.) *Kankogaku-heno-tobira (Door to Tourism Academy)* (pp. 36–50). Tokyo: Gakugei-shippan.

Sato, K. (2008b) Mass Tourism-no-jidai (Era of mass tourism). In M. Inoue (ed.) *Kankogaku-heno-tobira (Door to Tourism Academy)* (pp. 18–34). Tokyo: Gakugei-shippan.

Slow Food Japan (2010) On slow foods – Online document: http://www.slowfoodjapan.net/

Slow Life Centre (2011) Toshi to nosanson koryu: Slow Life Centre (Urban and country-side communication centre) – Online document: http://slowlife-c.com/index.html

Slow Life Japan (2010) Kankyu jizai no ikikata wo (Flexible use of slow and fast) – Online document: http://www.slowlife-japan.jp/

Smith, V. and Eadington, W. (eds) (1992) *Tourism Alternatives*. Philadelphia, PA: University of Pennsylvania Press.

Soshiroda, A. (ed) (2010) *Kankou machizukuri no marketing (Marketing for Community Development through Tourism)*. Tokyo: Gakugeisha.

Sugiyama, M. and Nobuoka, S. (2007) Slow food kara slow tourism he (From slow food to slow tourism). *Tokai Women's Junior College Kiyou* 33, 1–8.

T-Gate (2011) Tabihatsu – Online document: http://tabihatsu.jp/

Takai-Tokunaga, N. (2007) The dialectics of Japanese overseas tourists: Transformation in holiday making. *Tourism Review International* 11 (1), 67–83.

Takeuchi, D. (2005) Slow Tourism wo Motomete (In search of slow tourism) – Online document: http://rilc.forest.gifu-u.ac.jp/pdf/13slowlife.pdf

Team Sakimori (2010) Iki kasseika shudan gairyaku (About Team Sakimori: A group for the revitalisation of Iki Island). Unpublished document.

Teruoka, I. (1989) *Yutakasa no Joken (Conditions of Prosperity)*. Tokyo: Iwanami.

The Sloth Club (2010) On Sloth Club – Online document: http://www.sloth.gr.jp/

Uchida, K. (2006) Slow tourism no yukue (Future direction of slow tourism). In Ryoko-sakka no Kai (eds) *Slow Tourism no Tenbo (Perspectives on Slow Tourism)* (pp. 6–11). Tokyo: Gendai Ryoko Kenkyujo.

Upham, V. (1987) *Law and Social Change in Japan*. Boston, MA: University of Harvard Press.

Virilio, P. (1986) *Speed and Politics*. New York: Semiotext.

Watanabe, M. (1995) Kindai niokeru Nihon-jin no Shizen-kan: Seiyo tono Hikaku *nioite* (The Japanese view of nature in the modern era: A comparison with Westerners). In S. Ito (ed.) *Japanese View of Nature* (pp. 329–370). Tokyo: Kawade Publishers.

Wheeller, B. (2003) Alternative tourism – A deceptive ploy. In C. Cooper (ed.) *Classic Reviews in Tourism* (pp. 227–234). Clevedon: Channel View Publications.

Yamaguchi Prefecture (2007) Yamaguchi Slow Tourism: Promoting the Approach – Online document: http://www.yamaguchi-slow.jp/contents/about.html#houshin

Yamaguchi Prefecture (2010) Yamaguchi slow tourism web – Online document: http://www.yamaguchi-slow.jp/

Yamamoto, D. and Gill, A. (1999) Emerging trends in Japanese package tourism. *Journal of Travel Research* 38 (2), 134–143.

Yokohama, S. and Sakurai, T. (eds) (2006) *Potential of Social Capital for Community Development*. Tokyo: Asian Productivity Organisation.

14 Tribe Tourism: A Case Study of the Tribewanted Project on Vorovoro, Fiji

Dawn Gibson, Stephen Pratt and Apisalome Movono

The last 30 years has seen the phenomenal growth and diversification of the tourism industry to cater for the demands of the experienced and discerning traveller. Many consumers are no longer satisfied with traditional sun, sea and sand holidays. The motivation to experience nature and Indigenous cultures as part of the travel experience has increased (Poon, 1994; Swarbrooke & Horner, 1999). There is recognition that sustainable tourism should be concerned with quality over quantity. Quality relates to a deeper connection with fellow travellers, as well as the host community. A quality tourism experience can be difficult to achieve with over-full itineraries and a superficial interaction with the destination (Andereck *et al.*, 2006; Jennings, 2006). Often this involves running through a checklist of 'must-sees' and 'must-dos'.

As noted by Weiermair and Mathies (2004) slow tourism involves two principles – taking time and a connection to a geographical place. The attachment to a particular place is what others have called 'geotourism' (National Geographic Society, 2010), which encompasses not only the physical geography but also its culture, environment, heritage and the well-being of its residents. Thus, slow tourism involves considering the impact tourists have on the natural, cultural and economic environment. One of the key challenges facing tourists and the wider society is the creation of a more sustainable future. Innovative solutions to more sustainable and holistic living are going to be essential if people are to live within their means, to protect and respect the natural environment.

When a tourism operation actively seeks economic, environmental and sociocultural sustainability it aligns closely with the concept of slow tourism. 'Slow tourism' is an emerging concept which has been defined as providing 'an alternative to air and car travel, where people travel to destinations more slowly overland, stay longer and travel less' (Dickinson & Lumsdon,

2010: 1). This mode of travel covers the whole holiday and includes experiential factors like:

> the importance of the travel experience to, and within, a destination, engagement with the mode(s) of transport, associations with slow food and beverages, exploration of localities in relation to patrimony and culture and a slower pace and, what might best be described as, support for the environment. (Dickinson & Lumsdon, 2010: 2)

Whilst there are other definitions, several key themes come through in terms of what it is. Slow equates to quality time and an enhanced quality tourism experience. It involves a meaningful and engaging experience which is in tune with the cultural and physical environment (Dickinson & Lumsdon, 2010: 4).

This chapter reviews the sustainability of a unique eco-community on the island of Vorovoro in Northern Fiji. Termed 'Tribewanted', this unique tourism project provides insights into how alternative tourism projects can provide benefits for marginalised communities, and minimise negative impacts whilst still maintaining a unique visitor experience in line with emerging expectations of slow travel and experiences of place. It evaluates the Tribewanted project as a form of community-based volunteer tourism that fosters a more sustainable travel experience both in terms of the visitor and the Indigenous host community. This case study describes how the tourists go through a transformative experience, primarily through cultural immersion, to realise real and meaningful connections with people, places, culture, food, heritage and the environment. As a form of slow tourism, this tourism project on Vorovoro provides an avenue for close contact with local people, culture and heritage, emphasises the impact on the environment of all decisions and actions taken by the tourists and aims to be as authentic as possible. As will be discussed later, the experiences on the island have changed the way the tourists live after they returned home.

Tribewanted on Vorovoro

Vorovoro is a small island of roughly 100 acres located 45 minutes' boat ride from Labasa (Fiji's fourth largest town with a population of around 25,000 citizens) in Vanua Levu, which is Fiji's second largest island (Figure 14.1). The island is marked by high ridges running along the island like a backbone. In the valleys though, the flat land is suitable for crop cultivation and buildings. This flat land has rich loamy soil which is conducive for fruit and vegetable agriculture.

Tribewanted claims to be a sustainable development project whose aims are: community building, sustainable living and an adventurous experience.

Figure 14.1 Location of the Tribewanted project, Vorovoro, Fiji

Source: Europe Hotels, 2010.

Tribewanted is an alternative tourism experience whose tribe members fall within the backpacker/volunteer/adventure tourism market. It consists of an online community (The Tribe) of nomads, hunters, warriors and a chief who spend their days working together with Team Fiji to build the real space for the Tribewanted online community. The project has been described as MySpace with a real space; Second life with a real life; Facebook with actual face-to-face meetings (Keene, 2008). Founded by Ben Keene and Mark Bowness in 2006, the Tribewanted concept is where the global social media network links with a real world project, journey and events. Those interested in the project could sign up on the website as online tribe members. This enabled them to monitor developments that were occurring on the island via the website. It also enabled them to vote on all major decisions on the island. In September 2006, the first on-island tribe members arrived on Vorovoro to build a sustainable community. The initial lease on the island was for three years; however, the island has been closed since December 2010 and is to be re-opened in April 2012 under new lease arrangements. On-island members regularly updated the Tribewanted website with daily developments and interacted with the online tribe as to how things were going on the island. In 2008, Ben Keene published a book on his experiences of creating a global online network of like-minded travellers and an Indigenous Fijian community as they built a new life on Vorovoro.

Before the project commenced the island of Vorovoro was uninhabited. Hence with the introduction of tourists both positive and negative environmental consequences could be solely attributed to tourism. Tribe members were encouraged to see that, particularly on a tropical island, every action has a consequence well beyond the beach. Both the local community and tribe members were encouraged to continue sustainable living after they returned to their places of residence.

Methodology

Data in this chapter were collected in three ways. The first method used for primary data collection was in-person semi-structured interviews conducted among tribe members (tourists) and Team Fiji (the local community members who were employed on the project). The second method used to collect data was through personal observation, an 'immersion methodology' (Jarvis & Peel, 2010). Here the research team travelled to Vorovoro and stayed on the island for seven days in April 2010, experiencing what it is like to live as tribe members and participating in all the activities the tribe members did. A significant amount of public relations and media was generated at the launch of this project. For example, *National Geographic* ran a feature in the February 2007 issue and various articles about the project appeared in the UK's *Guardian* and *Observer* newspapers. Further, the BBC shot a five-part observational documentary in January 2008 called 'Paradise or Bust' and the Australian travel programme, 'Getaway', ran a segment on the project. With high interest in this project, the research team decided to assess Tribewanted's claims to be a unique travel experience.

The research team comprised a cross-section of age, gender and race (Indigenous Fijian, mixed-race Fijian/Rotuman and Caucasian Australian). This was important, especially in the context of Fiji where strict protocols exist regarding the interaction between men and women of different ages. For example, it is inappropriate for a young Fijian male to interview an older Fijian woman in a one-on-one situation. Even in a group setting gender roles are very specific and there is little interaction between non-family males and females. Further, conducting a formal 'interview' or conducting a 'focus group' among Fijian men to draw out responses to a semi-structured questionnaire is alien to this cohort. In an effort to please the researcher it is thought that Fijians will provide the answer that they think the researcher will want to hear rather than give their own response (Evening, 2000; Gibson, 2003). What is more natural in eliciting responses from this cohort is to talk informally while drinking *yaqona* (kava) around the *tanoa* (communal kava bowl). In this situation Fijians do not perceive they are being interviewed, are more open and provide responses voluntarily. Additionally, with an Indigenous Fijian in the research team questions were able to be translated and back-translated in the same conversation, hence clarifying any issues that did not translate well into English.

The third method used to collect data was via an online survey. This online survey was conducted among tribe members who had recently departed the island. All tribe members were sent an email thanking them for visiting the island and encouraging them to provide feedback on their experience on Vorovoro, what they enjoyed and what they needed to improve. At the end of May 2010, 205 tribe members had responded to the online survey. This represents a response rate of 19.8% as the project overall has had 1034 tourists visit the island. The average age of a visiting tribe member is 28 years, while the average length of stay is two weeks. UK residents comprise the majority of tribe members with 61% travelling to Tribewanted from the UK and a further 13% coming from the USA. The main source of information of how tribe members found out about the project was through public relations, such as television, newspapers and magazines (38%), while over a quarter (28%) of tribe members first found out about the project through word-of-mouth from family or friends.

The tribe members were encouraged to continue their relationship with other tribe members in several ways after they left the island. These included through Facebook, MySpace and Flickr, as well as on the tribewanted.com forum. They were also encouraged to share their experiences with other online tribe members through a personal blog. This online activity after the tribe members have returned to their place of residence enables the tourism operation to maintain a relationship with tourists as well as providing the most effective marketing tool: word of mouth or in this case 'word of web' (Weinberg & Davis, 2005).

Sociocultural Impacts of Tribewanted

In 2006, *Tui* Mali traditionally approached village elders to oversee the cultural and social exchange project Tribewanted. This was a chance to showcase the traditions and character of the people of Mali to the world. An important role for Team Fiji was *veiqaravi* 'looking after' the *vulagi* or 'visitors'. This role was entrusted to Team Fiji by *Tui* Mali and is upheld to this day. The cultural experience of the project, that is, to live as if in a traditional Fijian village with all its customs and protocols, is a large motivator for visiting the island. The young, predominantly Western visitors adopt the traditional dress codes of Indigenous Fijians by dressing conservatively. Women are expected to wear *sulus* (long wraparound printed skirts worn by men and women) to cover their legs and tops with short sleeves to cover their arms. Gone are the short skirts, shorts and bikini tops. Men are less restricted and can wear shorts during the day but are expected to wear a shirt or T-shirt and *sulu* in the evening.

In addition to being involved with 'karmic duties' (daily jobs) such as cooking, firewood-collecting, gardening and feeding the animals (pigs and chickens), tribe members also learn and perform *mekes* (Fijian dances), present *sevusevu* (meeting and greeting ceremonies), take Fijian language lessons and sit around the *tanoa* (kava bowl) in the evening chatting, drinking kava and singing. The type of tourism developed on Vorovoro allowed the revitalisation of cultural practices which without the Tribewanted project would have become extinct. A good example of this is the ancient practice of '*veitamaki*' or 'to announce oneself before entering a home or village', a practice which villagers had stopped using in the district, but is now practised everywhere on Mali because it was reintroduced at Vorovoro.

Tribe members who completed the online survey rated these cultural experiences highly. On a five-point Likert scale, tribe members rated the school visits an average of 4.65/5 and the learning and performing of *mekes* an average of 4.68/5. Similarly, tribe members rated the kava drinking sessions an average of 4.43/5 and the village visits 4.53/5. In addition to specific cultural activities, tribe members were also able to interact on a day-to-day basis living with the local Fijians (Team Fiji). Hence, activities such as snorkelling on the nearby reef and participating in a cross-island hike were not only physical challenges but also became ways of developing friendships and sharing traditional knowledge with tribe members. As with the cultural activities satisfaction with these activities was high. The reef trip was rated an average of 4.59/5 and the Four Peaks Hiking Challenge as 4.48/5.

The Tribe

Maffesoli (1996) discussed the concept of tribes as coherent subcultures or neo-tribes of society, e.g. youth culture, that form 'a series of temporal gatherings characterised by fluid boundaries and floating memberships' (Bennett, 1999: 600). Richards and Wilson (2004b) suggested that backpackers could be characterised as 'neo-tribes' (Maffesoli, 1996) of social groups, that bond together temporarily in new uncertain environments. Bennett (1999) argued that youth were more fleeting and random, and capable of belonging to a variety of groups or 'neo-tribes' outside of their family home environment. Studies show that modern youth are likely to move between a number of groups with different foci (Hottola, 2002, 2008; MacCannell, 1999). In the case of cultural experiences they may find themselves inadequately equipped to deal with the 'authentic other', so they move back to the familiar in groups like 'The Tribe' or familiar backpacker enclaves nearby. Shields (1992, in Bennett, 1999: 606) suggested that the different collective identities in modern consumer societies demonstrated the temporary nature of group identities in modern society 'as individuals continually move between different sites of collective expression and "reconstruct" themselves accordingly'. For example, young travellers visit and engage in a variety of destinations and projects that share different identities. The youth tourism market offers a variety of destinations and experiences that cater for the backpacker/volunteer/adventure tourism market. There are the 'party' resorts in the Yasawa and Mamanuca Island groups of Fiji and then projects like Tribewanted that offer volunteer/adventure/cultural experiences working together with local Indigenous communities.

The younger tribe members who were on the island at the same time as the researchers had chosen a variety of experiences during their extended world travel and seemed to feel at ease moving between the different groups. A friendly welcoming atmosphere was found throughout. Whilst Tribewanted may face challenges marketing to potential clients, it is obvious from the response of many of their guests who extend bookings that this form of cultural exchange is a unique multi-cultural experience missing from the hectic backpacker travel scene. There were many examples of people, such as Lisa from Canada, who had chosen to join Tribewanted at the end of her 'grand tour' and thus extend her stay. Another tourist (Jenny from England, who later adopted the Fijian name *bebe*, or 'butterfly') went for a month as a tribe member and ended up staying 10 as the *wavu* or bridge between the tribe and Team Fiji. Jenny and Jimmy Cahill from Indiana took their family and stayed for 10 weeks, then returned to lead the project for a year and are now considering staying even longer.

> Our time on Vorovoro has given us gifts that will be a part of us forever – we have experienced enduring and strong connections to ourselves, to each other, and to our goals and intentions for our family. We have formed relationships that will last the rest of our lives. The important lessons of slowing down, laughing much, and enjoying the moment we are in have become a part of us. (Jenny Cahill, 2010, www. tribewanted.com)

For Tribewanted participants, backpacker and volunteer tourism can mean extended periods of separation from home and friends, although today cell phones, emails and internet blogs have made communication easier (Cohen, 2004). Access to the internet is sporadic at Vorovoro but of little interest to most tribe members. The feeling of 'family' found at Vorovoro seems to offer tribe members a sense of belonging they have missed 'on the road'. Tribe members commented how at home they rarely stopped to spend time with family, share meals together and relax, as their lifestyles are just too busy. Life on Vorovoro is simple and uncomplicated. It is normal for guests to extend their stay for weeks or even months, and farewells at Vorovoro can become very emotional occasions. The following are representative of how tribe members feel about the unique experiences they shared at Vorovoro:

> Tribewanted was a total breath of fresh air for me having followed the well trodden backpacker route through Thailand, Australia, NZ etc. and was easily the most memorable and fulfilling week I had on the whole trip. The highlights of my week on Vorovoro are too many to mention but the visit to local school and the welcome we received will stay with me for a long time. I also loved being able to leave my mark on the island by building the showers and dining table during the first week of the project. I got a lot of satisfaction from that and I used the experiences and skills that I learnt on the island to bag me a job when I got home! (Nick Cresner, 2008, www.tribewanted.com)

> I never could have imagined how special and influential my stay on Vorovoro would be. What was it exactly that made it so hard to leave? Of course the beautiful warm weather and soft waves are exquisite, the food is excellent, and the activities are exhilarating, but it was leaving my new Tribewanted family that brought on tears during my final boat ride away from the village. Fijians are the warmest, most welcoming and accepting people I have come across to date. I want to say *Yadra Sia* to my family, and sometimes I do. There are times when I think about something I want to say to *Bebe*, Tomasi, Api, or Pupu, and I wish I were still on the island. (Lisa Costa, 2010, www.tribewanted.com)

The Tribewanted project is a unique example of slow tourism, which advocates a sustainable livelihood that makes a direct contribution to the local community and at the same time promotes the Indigenous Fijian village lifestyle which encourages participants to slow down and be on 'Fiji Time'. In this context 'Fiji Time' is taken to mean relaxation rather than lateness. Interestingly, many of the comments made by tribe members highlight the strong relationships they developed with the host Fijians in a short period of time. Discussion of a 'home away from home' emphasised their ready acceptance and inclusion into the 'Fijian' tribe. While tribe members were quickly made to feel welcome by Team Fiji, they recognised that it was the context of a relatively remote island that enabled them to bond so quickly. Conversely, the relationships developed within tribe members (tourists), while not adversary, were not as close, especially amongst male members of the tribe. This may have been as a result of the type of tourist who is coming to Vorovoro – those seeking a more unique individual experience, off the beaten (backpacker) track. Also, the very nature of the 'adventurous backpacker' and the 'global nomad', in search of authentic destinations untouched by other tourists, may result in them wishing to escape or avoid other travellers (Jarvis & Peel, 2010; Richards & Wilson, 2004a).

The search for the 'exotic other' or authentic cultural experience are common motivations amongst backpackers and adventure travellers, who hope to return home with a 'newfound sense of identity or increased cultural awareness' (Wilson & Ateljevic, 2008: 103). Sometimes the exposure to different cultures can exceed a person's ability to adapt and must be carefully managed (Hottola, 2008). This is where the voluntary nature of participating in Tribewanted activities is important. Although engaging in unique cultural experiences may buy prestige and enable the traveller to stand out from the masses (Welk, 2004), the cultural experience at Tribewanted is not always easy and changed many tribe members forever, making them rethink their values, attitudes and Western consumerist lifestyles (J. Cahill, pers. comm., 14 April 2010; I. Campbell, pers. comm., 24 November 2010). Wilson and Ateljevic (2008) observed similar reactions in female backpackers and travellers in India, Ethopia and Kenya. The women they interviewed discovered that cultural capital was not easily attained, and led them to question their materialistic lifestyles, become more aware of the social and cultural limitations and customs of destinations and acknowledge the differences between Western and local women.

Tribe Members' Environmental Footprint

Tribewanted has a commitment to be as environmentally sustainable as possible. It does this through its water usage, food production and consumption, waste management and energy usage. For many tribe members the sustainability aspect and attempting to live sustainably for an extended period of time on a Pacific island was a driving force of their stay.

Water usage

There is no natural source of fresh water on the island. Given that the majority of tribe members are from the UK, USA or Europe, the dry season on Vorovoro corresponds to the peak tourist season, and this puts a strain on the limited water resources available on the island. Rainwater is the only source of fresh water. This is collected on the roof of the pig pen and captured in several large water tanks. Tribe members were conscious of the limited amount of water available on the island and were content to use salt water to wash their hands and clothes. There is also a waterfall at one end of the island which runs down continuously from a natural dam above after it has rained. Otherwise tribe members shower using small buckets with a faucet which is connected to the rain water stored in three large tanks. It is significant that no water is used for toilets. The toilets on the island are compost toilets.

Food production and consumption

With nothing cultivated on the island, when the project began it was necessary to transport all food onto the island. One of the environmental initiatives enacted on the island was to attempt to be as self-sufficient as possible, at least with the produce that could be grown on the island. This involved both the supply of food production and the demand side of food consumption. On the supply-side, the decision was made to grow fruit and vegetables such as pawpaw, coconut and other garden vegetables. Free-range chickens were brought in to produce eggs and all of the fish were caught in nearby waters.

Many of the examples noted above were discussed amongst both the on-island and online communities and deliberate decisions were made for the benefit of the local economy and the desire to live more sustainably, often forsaking immediate self-gratification for the social good. Awareness of the economic, environmental and health impacts of food production and supply is raised when the tribe members are directly confronted with the conse-quences of their actions. This cognition function is often followed by a change in behaviour. A significant number of tourists reported taking more consideration when they purchase items at their local supermarkets and eliminate the use of plastic bags. Several tribe members reported changing to eco-friendly cleaning products, reducing their consumption of pre-packaged food to reduce their packaging waste further, growing their own fruit and vegetables in their garden/allotment and in general trying to live a simpler, healthier lifestyle. One visitor stated when they needed to purchase a new washing machine they bought one which is more eco-friendly, weighing the washing to save water. Ian Campbell, who came for a week and stayed seven months as the Tribewanted *wavu*, says:

> I have learned respect for the environment and other people, a sense of community, knowledge of Fijian culture and being able to live with what I need and not what I want. (I. Campbell, pers. comm., 25 November 2010)

Waste management

Despite an effort to source as much as possible from the island itself, to run a tourism project a certain amount of goods need to be brought onto the island. This produces waste. An analysis of the rubbish (calculated and weighed) found that of all the rubbish produced 60% is returned to the main-land (Labasa) dump for landfill (there is no landfill on the island). The remaining waste is retained for potential reuse. This consists primarily of glass bottles, plastic bottles and various plastic tubs. One example of the reuse of plastic tubs is as a receptacle for seedlings for starting new agricultural crops. Tribe members reported that research into recycling and sustainable practices opened their minds as to what they can do, how much they need to do, and how little they really need – spending too much time 'wasting and spending' instead of conserving, re-using and revitalising. The time on the island presented them with an opportunity to look at ways of living in the present moment with what they have, not what they would like to have, primarily enjoying the people and environment.

Tribe Members' Economic Contribution

Tribewanted has collectively invested over $FJ2 million ($US1.06 million) into the local economy in Northern Fiji. The project has generated 20 full-time jobs, fundraised for four villages and through the supplies and materials that tribe members bring it has supported the development of the local school and communities. The business model that Tribewanted uses seeks to minimise administration costs to contribute maximum benefits to the local community. Tribe members' fees, which average $US320 for a seven-night stay, include accommodation and all meals, cover the local experience and add to the development of the island. Seventy cents in the dollar is invested locally in land leases, development materials, equipment, employment, training, transport, food and insurance (Tribewanted, 2011). Another 20% is spent on commissions, marketing and public relations, while only 10% is spent on internet fees and web management and accounts.

Tribewanted as Slow Tourism

Slow tourism has been suggested as a conceptual framework which 'encompasses the whole holiday [and] ... suggests that engagement with people and place is an important theme' (Dickinson & Lumsdon, 2010: 88). As explained earlier, the rich cultural interaction that tribe members have with Team Fiji during the course of their daily lives on Vorovoro is a unique form of slow tourism. There are no roads, the only way to travel is on foot or by local boats. Trips to the mainland are kept to three times a week which minimises fuel use. However, the trip from Labasa town to Vorovoro down the Labasa river, lined with bat-filled old mangrove forests and then out into the open sea to Vorovoro is an experience in itself.

Socialisation is an important motivation for travel and although many backpackers say they are travelling alone or in pairs, in fact they are following a well-travelled road of young people meeting and forming impromptu groups during the course of their travels, and gathering in backpacker enclaves like the Yasawas in Fiji (Loker-Murphy, 1995; Maoz, 2007; Newlands, 2004). Tribewanted offers an alternative off-the-beaten track experience that also fulfils the travel motivations of meeting new people, making friends, sharing with people of similar interests, taking time out with family and friends, interacting with local cultures and directly contributing to the local community (Richards & Wilson, 2004a; Wearing & Neil, 2000). However, the cultural exchanges at Vorovoro take place over longer periods of time than the fleeting contact with locals provided by other tourism experiences, and are thus thought to be more meaningful and authentic, an important feature of backpacking and volunteerism (Brown, 2005; MacCannell, 1999; Richards & Wilson, 2004b). Throughout their stay on Vorovoro, tribe members are interacting with each other and sharing travel and life experiences, as well as with members of Team Fiji. Elders like *Tui* Mali, Pupu Epeli, Iliavi and *Ratu* Poasa have a lot of wisdom to share with many of the young tribe members. Life stories are shared on medicinal tours, while working in the *teitei* (gardens), on fishing trips, or sitting in the Grand *Bure* in the evening sharing a few bowls of kava. Tribe members immerse themselves in this unique cultural experience and take time out from their hectic Western lifestyles – this too is an aspect of slow tourism (Dickinson & Lumsdon, 2010).

Evidence suggests that the travel experience of this particular project has changed many tribe members' attitudes to sustainability and promoted more environmentally beneficial behaviour after the tribe member returned home (Tyer, 2010; Wilson & Ateljevic, 2008). For example, Helen Wolstenholme from Yorkshire in the United Kingdom stated that her stay at Tribewanted gave her 'an eye opening lesson in how wasteful we are as a world and what being part of a true community really feels like' (Tribewanted, 2011). There is a high likelihood to recommend Tribewanted to family and friends (92% of tribe members stated they would recommend Tribewanted) while 84% agreed that after leaving Tribewanted they wanted to make significant changes in their life. Tyer (2010) noted three areas where tribe members acted more environmentally responsibly: consumption; transport and wider community. More sustainable consumption involved the purchase of more biodegradable products such as detergent and washing powder as well as purchasing more local, seasonal and organic produce.

Travel for many youth travellers in the backpacker/volunteer tourism market is considered to be a rite of passage where they have the freedom to explore new experiences outside the boundaries of their home environment and provides a sense of achievement felt by young travellers on their return home (Brown, 2005; Cohen, 2004). Modern-day backpackers are increasingly engaging in alternative tourism and volunteering projects as part of their extended travel experience (Matthews, 2008). Many tribe members had followed the conventional backpacker routes through Australia, New Zealand

and Fiji and then elected to participate in the Tribewanted project to fulfil their altruistic motivations of engaging with and contributing to an Indigenous host community (Brown & Morrison, 2003). Whilst previous research has centred on the different motivations of backpacker and volunteer tourist experiences, emerging research has now identified overlapping similarities related to volunteering that suggests a more sustainable form of backpacker tourism is evolving that includes volunteering and alternative tourism projects like Tribewanted (Ooi & Laing, 2010).

Volunteer tourism is a combination of travel and voluntary work which provides opportunities to 'find yourself' through personal development as well as making positive contributions to the social, environmental and economic environments of local host communities (Broad, 2003; Brown, 2005; Dickinson & Lumsdon, 2010; Ooi & Laing, 2010; Raymond & Hall, 2008; Wearing, 2001). There is evidence to suggest that such experiences can lead to improved cross-cultural understanding between volunteers and hosts, where both the volunteer and the host benefit (Raymond & Hall, 2008). However, when volunteers assume the roles of 'expert' and 'teacher' negative neo-colonial stereotypes of 'them' and 'us' can arise (Simpson, 2004). Volunteering has the potential to be a social agent of change, correcting stereotypes and misconceptions of both host communities and fellow volunteers and transformative in changing individuals (Cohen, 2004; Lepp, 2008; Ooi & Laing, 2010). Ooi and Laing (2010: 195) discussed the potential of volunteer tourism to 'facilitate personal growth and self-awareness ... [enabling individuals] to explore and develop the "self" through their altruistic efforts'. For example, on their return home tribe members engaged in more environmentally friendly transport which involved more frequently walking or cycling for both commuting and recreational purposes. Tribe members impacted on their wider community by being a spokesperson for environmental awareness through the education of family and friends, communities and workplaces. This transfer of knowledge from the tourists to their peers is a form of vicarious learning through which sustainable environmental practices can be shared, without having to travel to the destination. However, as with previous studies, some found that those at home were uninterested in their experiences, or the increased confidence and self-esteem they had developed through what they perceived to be hard-earned 'cultural capital' (Wilson & Ateljevic, 2008).

Conclusion

The Tribewanted tourism experience contains many elements of slow tourism. The Fijian way of life and the concept of 'Fiji time' are synonymous with slow tourism. Tribe members (tourists) are encouraged to relax and refresh mind and body. They do this primarily through, not just contact, but, immersion in the South Pacific culture and heritage through genuine friendships with the Indigenous Fijian people. Feedback through the post-trip online survey suggests overall satisfaction with the experience is extremely high, particularly with the cultural elements of the tourism project. Given the interaction with the local community and the tourists, these cultural aspects are perceived as authentic. The food, as much as possible, is sourced from the local market, if not grown on the island itself. Imported foods are kept to a minimum. Furthermore, the tourism project tries to be as environmentally sustainable as possible making the most of the limited water supply and re-using and recycling where possible resulting in as small a carbon footprint as possible.

Tourism as it has been developed at Vorovoro has enabled Team Fiji to fulfil *Tui* Mali's dream of sharing their culture with the world, enabled them to earn a living and in many cases revive different aspects of Indigenous Fijian culture. The multi-cultural exchanges within the tribe and between the tribe and Team Fiji provide authentic cultural exchanges, whilst building a sustainable, self-sufficient eco-community. As discussed above, the benefits to the Mali community have been positive and the concept of sustainable living and minimising one's carbon footprint has even impacted on the tourists' lifestyles when returning home. Furthermore, this awareness and the experiences of the tribe and Team Fiji are communicated globally using social network media. Tribewanted is an eco-community that provides a novel cultural experience that not only fulfils travel motivations of meeting new people and socialising but enables meaningful interaction with the Indigenous Fijian villagers of the Mali region of Northern Viti Levu in Fiji.

References

Andereck, K., Bricker, K.S., Kerstetter, D. and Nickerson, N.P. (2006) Connecting experiences to quality: Understanding the meanings behind visitors' experiences. In G. Jennings and N.P. Nickerson (eds) *Quality Tourism Experiences* (pp. 81–98). Oxford: Elsevier Butterworth-Heinemann.

Bennett, A. (1999) Subcultures or neo-tribes? Rethinking the relationship between youth, style and musical taste. *Sociology* 33, 599–617.

Broad, S. (2003) Living the Thai life: A case study of volunteer tourism at the Gibbon Rehabilitation Project, Thailand. *Tourism Recreation Research* 28 (3), 57–62.

Brown, S. (2005) Travelling with a purpose: Understanding the motives and benefits of volunteer vacationers. *Current Issues in Tourism* 8 (6), 479–496.

Brown, S. and Morrison, A.M. (2003) Expanding volunteer vacation participation: An exploratory study on the mini-mission concept. *Tourism Recreation Research* 28 (3), 73–82.

Cohen, E. (2004) Backpacking: Diversity and change. In G. Richards and J. Wilson (eds) *The Global Nomad. Backpacker Travel in Theory and Practice* (pp. 43–59). Clevedon: Channel View Publications.

Dickinson, J. and Lumsdon, L. (2010) *Slow Travel and Tourism*. London: Earthscan.

Europe Hotels (2010) Fiji map – Online document: http://www.europehotels.gr/gr/exotic/images/fiji-map.gif

Evening, E.S.H. (2000) A case study investigation of the factors responsible for limiting the marketing exposure of small-scale village-based schemes in Fiji. Unpublished Masters thesis, Lincoln University, New Zealand.

Gibson, D. (2003) More than smiles: Employee empowerment facilitating the delivery of high quality consistent services in tourism and hospitality. Unpublished Masters thesis, University of the South Pacific, Suva, Fiji.

Hottola, P. (2002) Touristic encounters with the exotic west: Blondes on the screens and streets of India. *Tourism Recreation Research* 27 (1), 83–90.

Hottola, P. (2008) The social psychological interface of tourism and independent travel. In K. Hannam and I. Ateljevic (eds) *Backpacker Tourism: Concepts and Profiles* (pp. 26–37). Clevedon: Channel View Publications.

Jarvis, J. and Peel, V. (2010) Flashpacking in Fiji: Reframing the 'global nomad' in a developing destination. In K. Hannam and A. Diekmann (eds) *Beyond Backpacker Tourism: Mobilities and Experiences* (pp. 21–39). Bristol: Channel View Publications.

Jennings, G. (2006) Perspectives on quality tourism experiences: An introduction. In G. Jennings and N.P. Nickerson (eds) *Quality Tourism Experiences* (pp. 1–21). Oxford: Elsevier Butterworth-Heinemann.

Keene, B. (2008) *Tribe Wanted: My Adventure on Paradise or Bust*. London: Ebury Press.

Lepp, A. (2008) Discovering self and discovering others through the Taita Discovery Centre volunteer tourism programme, Kenya. In K.D. Lyons and S. Wearing (eds) *Journeys of Discovery in Volunteer Tourism* (pp. 86–100). Wallingford: CABI.

Loker-Murphy, L. and Pearce, P.L. (1995) Young budget travellers: Backpackers in Australia. *Annals of Tourism Research* 22 (3), 819–843.

MacCannell, E. (1999) *The Tourist: A New Theory of the Leisure Class* (3rd edn). Berkeley, CA: University of California Press.

Maffesoli, M. (1996) *The Time of Tribes: The Decline of Individualism in Mass Society*. London: Sage.

Maoz, D. (2007) Backpackers' motivations: The role of culture and nationality. *Annals of Tourism Research* 34 (1), 122–140.

Matthews, A. (2008) Negotiated selves: Exploring the impact of local-globalised interactions on young volunteer travellers. In K.D. Lyons and S. Wearing (eds) *Journeys of Discovery in Volunteer Tourism* (pp. 101–117). Wallingford: CABI.

National Geographic Society (2010) About geotourism – Online document: http://travel.nationalgeographic.com/travel/sustainable/about_geotourism.html

Newlands, K. (2004) Setting out on the road less travelled: A study of backpacker travel in New Zealand. In G. Richards and J. Wilson (eds) *The Global Nomad. Backpacker Travel in Theory and Practice* (pp. 217–236). Clevedon: Channel View Publications.

Ooi, N. and Laing, J.H. (2010) Backpacker tourism: Sustainable and purposeful? Investigating the overlap between backpacker tourism and volunteer tourism motivations. *Journal of Sustainable Tourism* 18 (2), 191–206.

Poon, A. (1994) The new tourism revolution. *Tourism Management* 15 (2), 91–92.

Raymond, E.M. and Hall, C.M. (2008) The development of cross-cultural mis(understanding) through volunteer tourism. *Journal of Sustainable Tourism* 16 (5), 530–543.

Richards, G. and Wilson, J. (2004a) Drifting towards the global nomad. In G. Richards and J. Wilson (eds) *The Global Nomad. Backpacker Travel in Theory and Practice* (pp. 3–13). Clevedon: Channel View Publications.

Richards, G. and Wilson, J. (2004b) The global nomad: Motivations and behaviour of independent travellers worldwide. In G. Richards and J. Wilson (eds) *The Global Nomad. Backpacker Travel in Theory and Practice* (pp. 14–39). Clevedon: Channel View Publications.

Simpson, K. (2004) 'Doing development': The gap year, volunteer-tourists and a popular practice of development. *Journal of International Development* 16 (5), 681–692.

Swarbrooke, J. and Horner, S. (1999) *Consumer Behaviour in Tourism*. Oxford: Butterworth-Heinemann.

Tribewanted (2011) http://www.tribewanted.com

Tyer, S. (2010) Can travel experiences impact on pro-environmental behaviour? Unpublished Masters thesis, The Graduate School of the Environment, The Centre for Alternative Technology, Machynlleth, Powys, Wales.

Wearing, S. (2001) *Volunteer Tourism: Experiences That Make a Difference*. Wallingford: CABI.

Wearing, S. and Neil, J. (2000) Refiguring self and identity through volunteer tourism. *Society and Leisure* 23, 389–419.

Weiermair, K. and Mathies, C. (2004) *The Tourism and Leisure Industry: Shaping the Future*. New York: The Haworth Press.

Weinberg, B.D. and Davis, L. (2005) Exploring the WOW in online-auction feedback. *Journal of Business Research* 58 (11), 1609–1621.

Welk, P. (2004) The beaten track: Anti-tourism as an element of backpacker identity construction. In G. Richards and J. Wilson (eds) *The Global Nomad. Backpacker Travel in Theory and Practice* (pp. 77–91). Clevedon: Channel View Publications.

Wilson, E. and Ateljevic, I. (2008) Challenging the 'tourist-other' dualism: Gender, backpackers and the embodiment of tourism research. In K. Hannam and I. Ateljevic (eds) *Backpacker Tourism: Concepts and Profiles* (pp. 95–110). Clevedon: Channel View Publications.

15 Slow Tourism Initiatives: An Exploratory Study of Dutch Lifestyles Entrepreneurs in France

Esther Groenendaal

This chapter reflects upon the results of a mixed method study on the values, motives and lifestyles of Dutch micro scale tourism entrepreneurs in rural areas in France and the relation to slow tourism. The study is on the one hand embedded in the theory of social movements and human worldviews since the 1960s, and on the other hand in the theory of the Cultural Creatives who are working to evolve a new set of values referring to issues that are close to their heart. With an eye on globalisation issues implicated in shifting human consciousness, theories of social movements were used to explore values, lifestyles and mindsets among voluntary emigrants within the European Union. A significant number of these voluntary emigrants turn towards micro or small tourism entrepreneurial concepts and appear to create a tourism demand in line with their personal values and worldviews.

The results of the research presented here reveal a diversification in slow tourism initiatives. The research involved an internet content analysis of 300 websites and weblogs, 204 questionnaires and 10 in-depth interviews among Dutch tourism entrepreneurs in France. I argue that this study shows evidence of a rise in slow tourism initiatives by lifestyle entrepreneurs, not only as an answer to global acceleration but also as an outcome of progress in human consciousness about people, planet and pace. My research shows evidence of global, local and personal consciousness with direct impacts on micro scale tourism supply. Moreover, the increasing migration figures within the European Union (EU) predict a growing group of voluntarily migrating individuals. Considering the EU's aim to integrate communities and facilitate cross-border actions, this trend has only just started and will most likely become increasingly important in the near future. In relation to this trend the chapter provides suggestions for future research related to slow tourism development in rural areas within the EU, specifically the relation between voluntary migration and slow tourism supply.

Conceptualising 'Slow'

The rising popularity and the meaning of the word 'slow' provides food for thought in relation to the high-speed developments our globalised world encounters. Contemporary theories of globalisation embody various concepts of speed such as acceleration, physical and digital movement, global interconnectedness (Eriksen, 2007) and new norms of inclusion (Held *et al.*, 1999; Munar, 2007). Slow, for that matter, stands in stark contrast to the conceptualisation of today's fast changing world. However, the contrary seems true since the revaluation of the word 'slow' appears to have emerged as a response to a speedy pace of life. Or, as Honoré (2005: 16) states: the existence of 'the slow movement overlaps with the anti-globalization crusade'. 'Slow' plays a significant role in the globalisation debate and it deserves to be analysed in depth.

The pace of life is considered a subjective and ideological concept that is associated with Western mindsets and modernity (Germann Molz, 2009). In terms of tourism and travel, speed and slowness are related to ideas about mobility and travels from A to B (Peeters, 2007). In a similar way to studies of sustainable tourism, the slow tourism literature draws upon disciplines that, by their nature, take on a mobility, travel or transportation perspective (Dickinson & Lumsdon, 2010). In contrast, the research in this chapter moves away from mobility and transportation; it emphasises the experience of slow *at* a tourism destination and the role tourism entrepreneurs play in the sensory perception of being somewhere. The objective of unravelling human experiences of slow tourism is to contribute to a more nuanced debate on the value of tourism as a global phenomenon for both the 'consumer' and the 'producer' of slow tourism experiences. After a discussion of the theoretical context, including the significance of social movements and Cultural Creatives, the chapter will analyse values and lifestyles of micro scale entrepreneurs who encourage slow tourism demand. Careful attention will be paid to the way time and pace is positively interpreted in the context of tourism experience and management. The purpose of this chapter is to examine how 'slow' can be made meaningful in the context of tourism experiences and to contribute to debates about the relationship between tourism supply, tourism demand and sense of time in an era of globalisation.

The Slow Movement as a Social Movement

Reshaping realities and shifts in human consciousness frequently go hand in hand with the emergence of social movements. Historically, social movements emerged out of class struggles aligned with working-class or agrarian-class interests. In the contemporary era, social movements have a more global character that extends beyond social class affiliation or struggle, for example the environmental movement or the fair trade movement. In their book about Cultural Creatives as a global social movement, Ray and Anderson (2000) explicate how people have departed from traditional and modern cultures to contrive new ways of life. Ray (1996) distinguished three different streams of cultural meanings and worldviews: the traditional stream, the modern stream and the transmodern stream. Ray extends the modern stream to cover the developments denoted by postmodernity.

The social movement that has derived from the transmodern stream, defined by Ray and Anderson as the Cultural Creatives, responds to the major changes our civilisation currently experiences. Cultural Creatives literally create a new culture, a new mindset. They are claimed to be a mixture of various movements that have existed since the 1960s, searching for humane methods of social transformation (Ray, 1996; Ray & Anderson, 2000). This unmistakable trend in our globalised societies implies a longing for consciousness of the self in a holistic world.

A closer look at the core values and lifestyles of Cultural Creatives indicates a great similarity to the basic principles and values of the slow movement. Ray and Anderson (2000: 37) argue that Cultural Creatives highly value the planet, identifying it as 'the holistic everything'. They care about relationships, peace, social justice and about self actualisation, spirituality and self-expression. Besides, they can be characterised as both inner-directed and socially concerned. The overall keyword of Cultural Creatives' values and lifestyles is 'consciousness': about the position of the self, the planet, a sustainable future and societal transformation. Cultural Creatives, who are largely middle class, have an ability to think outside the box, making them innovators able to create solid new ground to turn their values into a new way of life.

As a complementary crusade, the worldwide movement of slow finds its origins in challenging speed and living life consciously. Followers of the slow movement want to strike a balance between the modern and the traditional by determining a personal and suitable pace of life. They value slowing down the subjective speed of a globalising world and re-thinking the abstractness of time and space (Honoré, 2005; Germann Molz, 2009). The step from the slow movement to tourism is a small one as both phenomena are grounded in notions of movement, pace and experience.

The linkage between the core values of the slow movement and those of the Cultural Creatives as a transmodern culture will be revealed through the presentation of the results of an exploratory study among micro scale tourism entrepreneurs in France. An analysis of values and lifestyles will contribute to the exploration of how entrepreneurial initiatives have shaped a platform for the development of slow tourism demand.

The Case of Micro Scale Lifestyle Tourism Entrepreneurs

The main purpose of the study presented here (Groenendaal, 2009) is to investigate lifestyle tourism entrepreneurs in order to identify a shift towards slow tourism experiences in terms of tourism supply and demand. The larger aim of the study is to contribute to the debate about tourism as a platform for changing worldviews by acknowledging the relevancy of lifestyle tourism entrepreneurship in shaping slow tourism demand.

The study is exploratory, seeking to find out 'what is going on here'. The subjects of study are individuals who voluntarily emigrated from one wealthy EU country (the Netherlands) to another (France). The study explores how entrepreneurs make progress in the tourism setting of their choice by giving meaning to their situation and environment. This method holds an interpretive approach which entitles the researcher and the participants to interact with the goal of understanding the situation from the participant's perspective. Quantitative and qualitative data were gathered through questionnaires, interviews and observation to deepen understanding. The following research questions were addressed:

- What shifts in human consciousness trigger the longing for voluntary emigration?
- What values and preferred lifestyles do voluntary emigrants have?
- To what extent are these values and preferred lifestyles used as a concept for tourism entrepreneurship?
- How do these tourism entrepreneurial concepts stimulate the demand for slow tourism experiences?
- What role do the core values of the slow movement and of Cultural Creatives play with reference to the emerging slow tourism phenomenon?

Method, Sampling and Research Techniques

To explore the meaning of slow in relation to tourism, the researcher searched for individuals who deliberately favour the quality of their life by taking the step to emigrate voluntarily. Additionally, they initiated a micro scale tourism enterprise. With those criteria in mind the decision was made to conduct the study in France as this is the most popular destination for Dutch emigrants to permanently move to. (Germany and Belgium, which are generally popular among border migrants, were not taken into consideration.) Besides, for many years France has been a highly popular tourism destination for the Dutch. The participants were recruited from searches conducted through portals and web-rings on the internet. Over 300 websites and blogs of tourism entrepreneurs in France were studied in an attempt to trace people, locate them and assimilate their emigrational, entrepreneurial and personal stories. The websites were randomly selected through portals such as www.nederlanders.fr, www.logerenbijnederlanders.com and www. infofrankrijk.com. Another line of selection was to follow web-rings such as www.chambresdhotes.nl/webring/ where entrepreneurs recommend other entrepreneurs with identical core values and similar tourism concepts on their own websites. The web-ring was initiated by a network of Dutch entrepreneurs in various regions in France. The purpose of studying the sites and blogs was threefold. Primarily, it would allow me to identify Dutch lifestyle tourism entrepreneurs that permanently live in France.

Second, I would be able to specifically address this population for the research. And finally, it would provide me with prior knowledge about lifestyles, entrepreneurial concepts and values of the research population. All sites provided visual evidence of the estate, the atmosphere and its setting by means of photos and maps. After finalising this phase personal contacts were established in order to efficiently collect primary data. The research subjects that suited the criteria (permanent migration, Dutch native, tourism entrepreneur, residency in France) were addressed. A personalised mail survey with 20 questions was developed and piloted in conjunction with micro scale tourism entrepreneurs. The questions were derived from the literature and from the content of the studied websites. The survey was designed to measure both demographic aspects, such as age, gender, family life cycle status and education, as well as psychographic aspects, including lifestyle characteristics, life values and emigration motives. The goal of this part of the research was to identify a profile of the average Dutch tourism entrepreneur in France as a basis for further qualitative research. Respondents were asked to identify mobility motives, values and various types of lifestyle characteristics using a 5-point Likert-type scale ranging from 'not important' to 'very important' and from 'not applicable to me' to 'very applicable to me'. The survey was prepared, tested and mailed in the first quarter of 2009. Two hundred and four surveys were sent out with 101 surveys being returned, representing a response rate of 49.5%.

After the analysis of the returned questionnaires the survey instrument was supported by 10 semi-structured, in-depth interviews and observations of small tourism enterprises. The 10 participants were selected from the outcome of the quantitative survey. The sampling method for this part of the study is characterised as criterion-based sampling (Ponterotto & Grieger, 2007: 414) which enabled the selection of participants who had in-depth experience with both emigration and lifestyle tourism entrepreneurship. The selected entrepreneurs were visited by the researcher on their own premises over a period of three weeks in April 2009. The private domain, the environment, the tourism product and the community were observed and photographed. The 10 semi-structured interviews were aimed at revealing personal beliefs of the participants. The topic list contained aspects such as:

- Motives to leave the Netherlands.
- Motives to live and work in tourism in France.
- Quality of life issues.
- Sustainability issues.
- Lifestyles.
- Core values, the meaning of slow.
- Sense of community and integration.
- Emigration and mobility.
- Tourism entrepreneurship.

Research Findings

Findings illustrate how the participants construct slow lifestyles and give meaning to slow tourism in line with the principles and values of the slow movement. The interlocking of slow tourism, micro scale tourism entrepreneurship and voluntary emigration within the EU will be examined in the following analysis.

What shifts in human consciousness trigger the longing for voluntary emigration?

As a characteristic of a globalising world, individuals move easily across borders. Voluntary migration flourishes in Europe especially, where the meaning of borders has radically changed with the unfolding of the EU. In the Netherlands, emigration figures have risen spectacularly and the reasons behind contemporary emigration have changed dramatically. History shows that the Netherlands has seen considerable emigration for centuries. Zeegers *et al.* (2004) argue that it started in the 12th century when emigration was forced by poverty and floods. People from the Netherlands fled to Belgium and Germany. Overseas emigration originated in the 16th century when Dutch settlers started a widespread Dutch colonial empire. In the 17th century the Dutch attempted to migrate to various Dutch settlements. In the 19th century the emigration wave to the Dutch Indies took shape and emigration to America increased heavily after 1840. Up to 1915 the number of Dutch emigrants to North America increased to 250,000. The most recent emigration wave occurred in the years following World War II. In the 1950s 560,000 Dutch people migrated to intercontinental destinations in an attempt to leave their war-degraded, overpopulated home country behind (Nicolaas, 2006). Now, at the beginning of the 21st century, emigration from the Netherlands is considered high. In 2006 132,470 people left the country, of which nearly 80,000 were native Dutch (Henkens & van Dalen, 2008). In the years 2005 and 2006 the figure increased to such a degree that a departure surplus was identified. Not only was this the first time it had ever happened in the Netherlands, it was also a unique situation for a Western-European country (Henkens & van Dalen, 2008: 7; Nicolaas, 2006: 35).

More important in the context of this study, however, are the reasons and motives for departure. Although it can be claimed that 'the country has a high standard of living with one of the lowest unemployment rates in Europe' (van Dalen & Henkens, 2007: 45), emigration does remain an issue in the Netherlands. Studies by the Demographic Institute (NIDI) and the Central Bureau for Statistics (CBS) have been conducted to discover not only numbers of emigrants, but above all motivations for leaving. The current emigration wave is therefore a curious development that cannot only be explained through dominant economic indicators which were the forcing power in earlier periods of emigration. Social and personal indicators appear to play an increasingly powerful role. In the Netherlands motives for migration are predominantly led by characteristics in the public domain of society. Henkens and van Dalen (2006, 2008) extensively examined the perception of individuals' public and private lives and the impact of both on the decision to emigrate. In their study the authors define the private domain through variables that can be controlled by individuals. The public domain on the other hand, refers to variables that are outside people's control and can only be addressed by means of collective action. Figure 15.1 shows evidence of the perception of the quality of the public and private domain in the Netherlands judged by Dutch individuals.

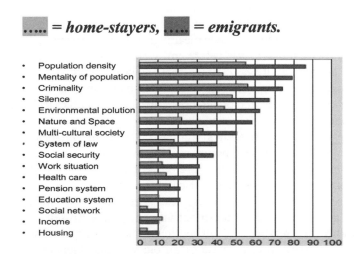

Figure 15.1 Quality of public and private domain in the Netherlands judged by emigrants and 'home-stayers'

Source: van Dalen *et al.* (2008).

Evaluating the quality of the public domain involves institutions and issues such as societal problems, welfare state issues and environmental quality. The private domain involves human capital, social networks and personal features. Van Dalen and Henkens (2007: 55) conclude that 'severe neglect of the public domain appreciably increases the pressure to emigrate'. This directly derives from their discovery that 'opinions regarding environmental pressure such as lack of space, noisiness and population density play a significant role in emigration decision making processes'. What emerges from this examination is that people increasingly gain awareness of what is happening in the public environment and how that influences their perceived quality of life. The reluctance against the density of the population and the derivative feeling of lacking space, as well as the negative perception of the mentality of that population, triggers people's antipathy towards society. Scarce phenomena such as silence, space, nature and cleanliness are highly valued by native Dutch emigrants.

What values and preferred lifestyles do voluntary emigrants have?

To trace the *core values* of the population under study, respondents were asked to identify the level of importance of various motives to move on a 5-point Likert-scale. Ten distinguished values and motives were derived from the literature (Henkens & van Dalen, 2006; van Dalen & Henkens, 2007; van Dalen *et al.*, 2005; van Dalen *et al.*, 2008) as well as from the website and blog analysis. The results of the survey indicate that the three major motives for people to voluntarily move are 'need for a better quality of life', 'need for rest and space' and 'longing for the countryside'. Seventy-eight percent labelled the 'need for a better quality of life' as an important or very important motive to move. The results of the interview sessions underline this point of view. Quality of life was conceptualised and made meaningful by the participants. Individuals critically question life as it was before emigration and life how they wished it to be after emigration. It was claimed that the perceived quality of life has in fact increased since emigrants took the step to move and set up their lives in line with their core values and preferences. Participants claim to feel happier and more satisfied with life. Moreover, they have a strong feeling of being financially independent which allows feelings of freedom and peace of mind. Economic wealth in terms of money making or business growth is, for most, of inferior importance. It is not interpreted as an indicator of quality of life.

The tourism businesses that participants created are in all cases a reflection of core values and chosen lifestyles. These businesses include accommodation and indoor and outdoor activities in line with self-actualisation and harmony. Key features are authentic and natural types of accommodation such as a watermill, a castle or an old vicarage. Generally speaking, entrepreneurs claim to be highly involved with local communities and local architecture. It is not uncommon that typical and meaningful objects are bought and restored to maintain the traditional feel of the countryside.

Besides, entrepreneurs make efforts to become engaged in local politics through membership of political parties, and become sponsors of sports clubs and other social events. Another significant feature of the migrated tourism entrepreneur is the sense of positive integration with host, guest and community. This leads to an unmistakable foundation for the co-creation of experiences by all three units and a tenor to a suitable pace of life.

The second highly motivational factor for individuals to move away from the Netherlands was 'need for rest and space'. Seventy-five percent labelled the 'need for rest and space' as an important or very important motive to move. This suggests that the level of unrest in the public domain of the Netherlands and the feeling of having too little space for oneself are prevailing. The Netherlands has a population density three times that of France. The quest for rest and space incorporates a 'longing for the countryside' and longing for valuable time. It is a core value of the participants which is directly translated to their conceptualisation of tourism.

With respect to *lifestyles*, information was gathered about the respondents' styles of living. The survey question about the characterisation of people's lifestyles asked individuals to identify to which level – ranging from 'not applicable' to 'very applicable' – 10 dimensions would apply to them on a 5-point Likert-scale. The results of the survey reveal that the 'feeling of freedom', 'sense of joy' and 'environmental consciousness' are the three dimensions the respondents feel most attracted to. Seventy-one percent labelled 'feeling of freedom' as the most important lifestyle characterisation, claiming it to be 'very applicable' to the person they feel they are. Sixty-three percent classified 'sense of joy' as the indicator of who they are. The meanings respondents give to the concepts of 'freedom' and 'joy' are again usually referred to as quality of life, high level of consciousness and suitable pace. 'Environmental consciousness' is the third most mentioned characteristic of participants' lifestyles. Environmental awareness among the participants does, however, not always result in climate neutral measurements since these are generally claimed to be 'too expensive to adopt'. However, creative environmental initiatives on a small scale are more common. Examples of such are visual adaptation to the environment when building new objects, the reuse of materials and the growing of own vegetables and fruits. A significant detail in this context is that nearly all entrepreneurs report about these initiatives on their websites and blogs. Guests and community members are informed about and involved in the environmental ambitions of the entrepreneurs.

How Values and Preferred Lifestyles Inform Tourism Entrepreneurship

The core values and preferred lifestyles of the tourism entrepreneurs under study form a significant foundation for the shape of the tourism supply and tourism experience. The sense of 'slow' in relation to tourism will be underlined in the following cluster of quotes, derived from in-depth interviews conducted with micro scale tourism suppliers in the context of life-style entrepreneurship. The following quotes identify personal core values the participants used as a thorough grounding for their tourism enterprise.

- Joke and Everard lived in different African countries and worked with people with various values and life priorities. They are well travelled, open-minded individuals with a great ability to see life in perspective. They value being broad-minded and they like to think global. 'We lived in Burkina Faso, Nigeria, Senegal and Ghana. Our son moved 12 times by the time he was 6. So when we came to France in 1999, we thought we'd slow down the moving pace a little. So we created the world in our rooms. We have an African room, a French room, a Japanese room …'.
- Guido is an optimistic person who values the chances he gets in life. He shows respect for the sphere of romanticism and nostalgia that adheres to the old railway station he bought. He exposed the old, yellow hotel pictures he found, reused the old furniture and is set to rebuild the hotel's history. 'Life is full of chances. And I truly felt I should take this chance and go for the railway hotel. I was not searching for a challenge; I just thought "wow!" I took on the chance, full of optimism, without considering any negative aspects. I believe in that method.'
- Corrie and Ron came to France to find the slow pace and simplicity. They value the peace of the countryside, the outdoor life and their animals. Corrie says: 'The goats, our beautiful rooster with his hens … the animals give me a peaceful feeling. So do the garden, the estate, the outdoors … I often realize how rich we are. This simple lifestyle is what I wish to share.'

Entrepreneurial Concepts and the Demand for Slow Tourism Experiences

Where the leitmotiv for most large and medium scaled enterprises is economic; small and micro sized enterprises are generally driven by non-economic, social and personal motives (Ateljevic, 2009; Ateljevic & Doorne, 2000; Shaw, 2005; Shaw & Williams, 2004). As a result of an increase in evidence indicating that lifestyle factors and non-economic motives are significant, it can be argued that a respectable and growing number of micro scale tourism entrepreneurs are in fact able and willing to create a tourism demand in line with their own core values. The rather non-commercial but personal approach is an answer to the quest for 'slow' in both supply and demand. It is interesting that every one of the participants runs his or her tourism enterprise in harmony with personal values. Although some are slightly more economically oriented than others, most are purely interested in sharing life values and lifestyles with guests. Not one neglects lifestyles and values in their tourism enterprise. 'Slow', in key terms of pace, joy and personal, as well as sense of community, sense of connection and sense of nature, reflects both the mindset of tourism suppliers and consumers. Illustrations of such are Janneke and Serge who have no economically oriented entrepreneurial ambitions at all. They entirely focus on a slow pace lifestyle which is what they offer as a tourism product, thus creating tourism demand:

> It is simple: we have no intention to adapt our level of supply. We do not offer guests what we ourselves do not demand. We are content with little, so we won't, – and now I exaggerate – install a bubble bath for the sake of tourism. We enjoy receiving guests here, but we won't deny our own ideas and mind-sets just to attract more guests and make money . . . . Guests that arrive here in a four-wheel drive for an unlimited holiday experience . . . . No, that does not fit. We like to have hikers and cyclists here. That suits our state of mind.

The Principles of Cultural Creatives Inform Slow Tourism

Since Cultural Creatives literally create a new culture and a new mindset (Ray & Anderson, 2000), and supporters of the slow movement value slowing down the subjective speed of a globalising world and re-thinking the abstractness of time and space (Honoré, 2005; Germann Molz, 2009), both phenomena may well be considered to make up the initial concept of slow tourism. Slow tourism is an emotional experience of life at a destination away from home for both the tourism supplier and consumer. It implies a shift from a fast, collective, general pace of everyday life to a slow, individual, conscious pace at the tourism destination. Participants of this study also mentioned the change in their existence, from accepting life as it is to questioning and challenging it and thus become more conscious of it. This concept is richly adapted to the tourism products most tourism entrepreneurs provide. This small-scale tourism supply creates a steady demand for the slow experience of time, space, nature and humanity. It seems valuable to seriously consider the development of slow movements and the growing importance of Cultural Creative mindsets when re-thinking tourism supply, demand and management.

Conclusion

The research presented in this chapter draws the contours of tourism as a platform for changing worldviews and shifts in consciousness. Mixed-method research among micro scale lifestyle entrepreneurs in France reveals that slow tourism is about people who wish to make a shift towards better understanding their personal world and adopting their own subjective concept of time to explore and recognise. Tourism is essential for that, given the fact that tourism implies 'being away from home'. The act of 'being away' provides new insights and a higher level of consciousness towards places, nature, culture and the self in a holistic world.

This chapter further adds to the debate on the role of tourism entrepreneurs in giving shape to the concept of slow tourism. In line with the findings of the research, I argue that the role of entrepreneurs in shaping slow tourism demand is a significant one, given the fact that most micro scale entrepreneurs as identified in this study are not driven by economic motivators but by consciously shaped personal values, based on the essence of life, time and space. As a result of their individual motivators and uncomplicated approach to tourism, entrepreneurs encourage the consumer to stay in one place rather than to travel around at high speed.

From this study two more research areas emerge. First, more understanding is needed of the activities and mindsets of consumers and local communities stemming from what is suggested here as slow tourism initiatives of tourism entrepreneurs. Although it appears as if tourism producers and consumers create slow tourism experiences together, empirical evidence is still lacking. Second, in order to get an extended insight into the dimension of the slow tourism phenomenon, further research on value-based entrepreneurial tourism concepts that create slow tourism demand is needed throughout the EU. Bearing the significant migration figures within the EU in mind, further research is necessary to improve our overall understanding of slow tourism.

References

Ateljevic, I. (2009) Trans-modernity – Remaking our (tourism) world? In J. Tribe (ed.) *Philosophical Issues of Tourism* (pp. 278–300). Bristol: Channel View Publications.

Ateljevic, I. and Doorne, S. (2000) 'Staying within the fence': Lifestyle entrepreneurship in tourism. *Journal of Sustainable Tourism* 8 (5), 378–392.

Dickinson, J. and Lumsdon, L. (2010) *Slow Travel and Tourism*. London: Earthscan.

Eriksen, T.H. (2007) *Globalization, The Key Concepts*. Oxford: Berg Publishers.

Germann Molz, J. (2009) Representing pace in tourism mobilities: Staycations, Slow Travel and The Amazing Race. *Journal of Tourism and Cultural Change* 7 (4), 270–286.

Groenendaal, E. (2009) Transmodern tourism entrepreneurship, an exploratory study on Dutch micro scale entrepreneurs in France. MSc thesis, Wageningen University.

Held, D., McGrew, A., Goldblatt, D. and Perraton, J. (1999) *Global Transformations: Politics, Economics and Culture*. Chicago, IL: Stanford University Press.

Henkens, K. and van Dalen, H. (2006) When the quality of a nation triggers emigration – Online document: http://publishing.eur.nl/ir/repub/asset/7600/2006-0261.pdf

Henkens, K. and van Dalen, H. (2008) Weg uit Nederland. Emigratie aan het begin van de 21ᵉ eeuw [Moving away from the Netherlands. Emigration at the start of the 21st century] – Online document: http://www.nidi.knaw.nl/en/output/reports/nidi-report-75.pdf/nidi-report-75.pdf

Honoré, C. (2005) *In Praise of Slow*. London: Orion Books.

Munar, A.M. (2007) Rethinking globalization theory in tourism. *Tourism, Culture and Communication* 7, 99–115.

Nicolaas, H. (2006) Nederland: Van immigratie- naar emigratieland? [The Netherlands: from a country of immigration to one of emigration]. *BevolkingsTrends CBS* 54, 33–40 – Online document: http://www.cbs.nl/NR/rdonlyres/DC8EE5A0- EA8A-40AC-A7AF-1EAD881747A0/0/2006k2b15p33art.pdf

Peeters, P. (2007) Tourism and climate change mitigation: methods, greenhouse gas reductions and policies (NHTV Academic Studies 6). NHTV Breda University of Applied Sciences.

Ponterotto, J. and Grieger, I. (2007) Effectively communicating qualitative research. *The Counselling Psychologist* 35 (3), 404–430.

Ray, P. (1996) The rise of integral culture. *Noetic Sciences Review* 37, 4–16.

Ray, P. and Anderson, S. (2000) *The Cultural Creatives: How 50 Million People Are Changing the World*. New York: Harmony Books.

Shaw, G. (2005) Entrepreneurial cultures and small business enterprises in tourism. In A. Lew, C.M. Hall and A.M. Williams (eds) *A Companion to Tourism* (2nd edn) (pp. 122–134). Oxford: Blackwell.

Shaw, G. and Williams, A.M. (2004) From lifestyle consumption to lifestyle production: Changing patterns of tourism entrepreneurship. In R. Thomas (ed.) *Small Firms in Tourism: International Perspectives* (pp. 99–113). Oxford: Elsevier.

Van Dalen, H. and Henkens, K. (2007) Longing for the good life: Understanding emigration from a high-income country. *Population and Development Review* 33 (1), 37–65.

Van Dalen, H., Henkens, K. and Bekke, S. (2005) Emigratie van Nederlanders: Geprikkeld door de bevolkingsdruk [Emigration of the Dutch: Triggered by population density]. *Demos bulletin voor Bevolking en Samenleving, Nederlands Interdisciplinair Demografisch Instituut* 21, 25–31.

Van Dalen, H., Henkens, K. and Nicolaas, H. (2008) Emigratie: de Spiegel van Hollands ongenoegen [Emigration: the mirror of Dutch dissatisfaction]. *BevolkingsTrends Nederlands Interdisciplinair Demografisch Instituut/CBS* – Online document: http://www.cbs.nl/NR/rdonlyres/0DCB 583A-8479-4B3E-B107-28A05E0747BA/0/2008k1b15p32art.pdf

Zeegers, M., Poppel, F. van, Vlietinck, R., Spruijt, L. and Ostrer, H. (2004) Founder mutations among the Dutch. *European Journal of Human Genetics* 12, 591–600.

16 Slow Travel and Indian Culture: Philosophical and Practical Aspects

Sagar Singh

In the anthropology of tourism, travel has been examined from various viewpoints, not merely for seeking its genesis, sustainability and status in the anthropological hierarchy of values, but also from the conceptual angle. Thus, tourism has been considered negatively as hedonistic play and subsumed by the term *Homo Ludus* (Graburn, 1983), a form of imperialism and neo-colonialism (Kobasic, 1996; Nash, 1989), or liminoid and frivolous activity (cf. Graburn, 1983). However, tourism has also been considered positively as a 'sacred journey' (Graburn, 1989; Timothy & Olsen, 2006). Most tourism anthropologists have seen tourism as a phenomenon that creates more imbalances and inequity than as a social process that has many positive sides to it. There has been one exception in the only form of travel that has traditionally been considered value-laden and 'slow' – pilgrimage. However, there are many other forms of tourism that have been considered progressive and positive, such as ecotourism (e.g. Buckley, 2003; Fennell, 1999) and volunteer tourism (Wearing, 2001), once their implications for society, ecology and other cultural and natural resources are identified. By explaining the importance of traditional travel in its Indian context, this chapter will examine the philosophical and pragmatic roots of slow travel. It will also explore the possible implications for forms of domestic (Indian) and international tourism that need to be addressed if travel is to be sustainable in all senses of the word – economic, social, cultural, ecological and political. Although research on the political sustainability of tourism is as yet an emerging field of study, this chapter will argue that slow travel generates sympathy, empathy and understanding and is, therefore, a process that can create peace. By focusing on traditional travel in India this chapter will also reveal how slow travel needs to be managed and well-planned (Singh, 2002, 2011a).

Slow travel is travel that is less energy intensive, has a smaller carbon or ecological footprint (like ideal ecotourism) and allows tourists to enjoy their experiences to a greater degree and extent by valuing socialisation, a greater appreciation of not only other cultures, peoples and their customs, but also scenery and landscapes. These slow qualities can contribute to feelings of harmony in the minds of tourists and hosts (Heitmann *et al.*, 2011). Through slow experiences, tourists and hosts have the opportunity for positive encounters that can facilitate better relationships between strangers in ways that also benefit local economies and uphold ecological values. By definition, therefore, slow travel can be more sustainable for the natural and cultural environments, tourism suppliers and tourists themselves (by increasing perceptual carrying capacity) (Singh, 2011b). In this way, slow tourism does not create a 'cognitive jetlag' among tourists; that is, the cognitive dissonance-cum-cultural shock that tourists often feel when transported quickly to far-off places that have a strikingly different culture and an incomprehensible language. This chapter will also examine the ramifications of slow travel for the continuation of promising but less popular forms of tourism in India, such as yoga tourism.

Background: Slow Travel and Indian Society and Culture

Although slow travel appears to be a newly-emerging phenomenon in the West, the societal structure of India was supported by processes that encouraged slow travel over the past 2000 years. At the time of the Buddha (circa 620 years before the common era, BCE), for instance, India had well-developed roads, inns, trees and wells along the routes that catered to pilgrims and slow travellers (Singh, 2009). These slow travellers mostly consisted of people exploring the cultural variety of the land or seeking education through slow travel (termed *deshatan* in Sanskrit) or completing the journey of their life through local, regional and country-wide pilgrimage (termed *teerthatan*). This continued in the beginning of the first millennium and gained more popularity in the 9th and 10th centuries of the common era (CE), after the establishment of pilgrimage circuits and the pan-India pilgrimage centres by Adi Shankaracharya, who contributed to uniting the two main sects of Hindus, Vaishnava and Shaiva, in philosophy and ritual (Kaur, 1985; Singh, 2003).

After India's Independence in 1947, the new fast means of transport, especially since the 1950s, seem to have spelled a decline in slow travel (Singh, 2002), but there are signs that both Indians and foreign tourists visiting India are taking to slow travel again with much enthusiasm. In several visits to the Western Indian Himalayas in the past 30 years, it was noticed that the number of pilgrims and trekkers has grown tremendously, with the Kedarnath route alone hosting 129,230 travellers in 1996 as compared to only

92,218 in 1976 (Singh, 2004a). This has been accompanied by a growth in ecotourism (Singh, 2004b), volunteer tourism (Singh, 2005), mountain-trek tourism and pilgrimages that involve strenuous walking. Even if the traveller reaches the starting point of the journey by modern road/rail/air transport, the idea of earning religious and cultural merit – that is, accumulating social, cultural and religious capital – spurs the tourist to undertake the arduous journey on foot. In this way, tradition mixes with modernity, a sure sign that travel had, and will continue to have, a significant role in maintaining the societal structure of India, where emphasis on doing things slowly was part of tradition.

How slow travel in India survived can be seen in the way Indian societal structure was shaped. Societal structure is a combination of the various social structures that are found in large societies or civilisations, such as religious structure, economic structure, political structure, kinship structure and so on. The patterns in which social relationships are organised are reinforced through a common language, religion, cultural universals such as music, social and religious festivals, and other institutionalised ways of behaving towards, in respect of, and with respect *for* the other (to paraphrase Nadel, 1942). Travel is an important form of communication, after language, which all civilisations require in order to survive.

The fact that needs to be emphasised is that Indian society today is more of a civilisation than a simple nation-state, and that it is, truly speaking, *Indian* civilisation, rather than Hindu civilisation. Various social and religious communities have contributed to its history, which, indeed, are as varied as those that exist in the world: Hindu, Muslim, Sikh, Christian, Buddhist, Jain, Ba'hai, Zoroastrian (Pharisee), Jewish and even African communities. The latter lead an independent existence on the west coast, in Gujarat and elsewhere. It should be noted that all these religious communities and the various sects that constitute them have co-existed, largely peacefully, for hundreds or thousands of years in India (Singh, 2011a). For example, Christians arrived in India a few years after Jesus Christ's death (St Thomas arrived in India circa 34 CE) (Basham, 1978). Zoroastrians arrived from Persia and settled in India around the same time. The present day Christians in India, however, are mostly converts that took to Christianity after the arrival of the British in the 17th century. Similarly, Muslims (Arabs) arrived and settled in India not along with, or after, the Moghuls in the 12th century CE, but around 742 CE, that is, around the time Prophet Mohammad was alive (Basham, 1978; Sharma, 1979, 1987).

As in Renaissance Europe and later (the grand tour), travelling was a means of education and socialisation with the richness of other cultures. During the time of the Buddha there were many roads, inns and famous old world universities (such as Nalanda in modern-day Bihar state and Takshila in the west of the Indian subcontinent) to which people travelled from far away to become enculturated. Travelling was a means of understanding and practicing religion and many followers of the Buddha became travelling mendicants and sages. Enculturation of common folk was also made possible by slow travelling minstrels and drama troupes (*nautanki* in north India and *jatra* – literally 'travellers' – in Bengal) that rendered ancient and folk versions of epics like the *Ramayana* and the *Mahabharata* (Bose, 1967). Thus, folk traditions and oral narratives (a number of which still exist), were complementary to and/or competed with Indian high culture, creating a rich diversity that was instrumental in creating travel demand. It was and is a resource for cultural tourism and a means of accumulating social and cultural capital; that is, the advantage that accrues to people of a society who are more cultured and who are more social, with resulting higher social status.

Long after the time of the Buddha, in the 16th century CE, the founder of the Sikh religion, Guru Nanak, upheld and reinforced this age-old tradition of travelling slowly and far and wide in order to understand religion and its essence, spirituality (Puri, 1982). People who travelled for religious purposes and pilgrims interacted with people from diverse social and ethnic backgrounds on the way, and interacted with more people than modern day travellers, since they travelled slowly, stopping at many places, thus leading to the formation and continuance of Indian societal structure. Slow travel in India, therefore, survived not only because many people were poor and could not afford fast travel (which now they can, but many still do not choose it), but also because it was essential for Indian societal structure, and essential for the personalities and resultant choice of mode of transport of rural people in India. This way of thinking, however, is not restricted to India, but to so many other peasant societies across the world (Wolf, 1966). Some anthropologists have termed slow and plodding man *Homo Faber* (Ardener, 1972).

Latest reports from the Ministry of Tourism, Government of India, confirm that pilgrimage still outweighs mass tourism by far (Times News Network, 2010). Among Sikhs, there are many shrines and temples in north India, but they either choose 'mixed' travel (utilising both fast modes of transport and relatively slow ones, say, to the most important shrine, the Golden Temple in Amritsar, Punjab and Patna Sahib in Patna, Bihar) or slow travel (to many other Sikh shrines, such as in Kurukshetra in the state of Haryana, Paonta Sahib in mountainous Himachal Pradesh, Nanakmata in Uttar Pradesh and Hemkund Sahib in the mountain state of Uttarakhand; to which latter pilgrims can go either on foot or by mule, the former being preferred by pilgrims or slow travellers). Lastly, one very important aspect of Indian slow travel, as epitomised in Hindu values, is non-violence towards animals and plants, and slow travellers do not (and cannot, since along such routes it is banned) eat meat, take alcoholic drinks or cut trees (Singh, 2002). In fact, the Indian tradition of slow travel and pilgrimage even exhorted the planting of trees along the way, and thus was beneficial for ecology (Singh, 1986). These values were made effective or translated into action by beliefs that whoever planted trees along the route would get *punya* (material or religious merit) and *moksha* (spiritual merit leading to freedom from the cycle of birth and death) (Singh, 2009). To explain this we turn to Indian philosophy and travel.

Slow Travel, Indian Culture and Philosophy

The most decisive influence on Indian (Hindu) culture was not that of the *Vedas*, the most ancient Hindu scriptures, but that of the *Geeta* (written by Lord Krishna sometime before the birth of Christ) and the *Puranas* (written between 200 CE and 500 CE) (Sharma, 1987). The *Geeta* explains yogic philosophy in as simple a manner as possible and, although considered an appendage of the epic, the *Mahabharata* (written by Veda Vyas, sometime between 200 BCE and 200 CE), stands out from the confused story of the latter. There are various yogic philosophies elaborated in the collective, known as Sankhya philosophy, or in the *Geeta*, which latter is said to be the fifth *Veda* (the other four, in order of importance as well as age, being *Rig Veda*, *Yajuh Veda*, *Atharva Veda* and *Sam Veda*) (Sharma, 1987). As a result, yogic philosophy as explained in the *Geeta* is also known as Vedanta philosophy (literally, the philosophy at the end of the Vedas: 'Veda' plus 'anta' or end). The *Geeta* explains three main yogic philosophies: Gyaan Yoga (yoga of knowledge and the intellect), Bhakti Yoga (yoga of divine love) and Karma Yoga (yoga of actions and deeds). This is different from the physical postures or yogic *asana*s explained by another great ancient Indian, Patanjali. However, even these physical postures have to be attained slowly and with control over the breath (also slow) and are precursors of control over the mind and, therefore, meditation. The actual meaning of yoga, often misunderstood by Westerners (due to misunderstanding by many Indians), has nothing to do with *asana*s or sex, but 'union with God' (Misra, 1976; Swami Ajaya *et al.*, 1979; Swami Rama, 1975). *Asana*s and meditation, according to yogic philosophy, have to be undertaken slowly (except when you are in the thick of battle, as Lord Krishna explains to the warrior, Arjuna, in the *Mahabharata*). The *Geeta* lays emphasis on *karma* and right action.

Perhaps the most difficult thing to explain in the world of Indian philosophy, including Dwaita (dualistic) and Adwaita (monistic) philosophy, is *karma*. *Karma* is usually translated as 'action' or 'deed', but there are times, as the *Geeta* says, when you are being guided by a guru and you have to take action without thinking of its consequences. Travel (tourism and pilgrimage) are also *karma*s but are considered 'good *karma*s' when seen in the light of the purpose of travel. In ancient as well as modern India, travel was and is undertaken for pleasure and trade, as well as for gaining religious or spiritual merit, or *punya* and *sanskara/moksha*, respectively. *Punya* can take the form of giving alms, clothes and food to the poor, helping the needy without any consideration of monetary gain, and has to do with the material world. But *sanskara* and the ultimate spiritual goal, *moksha* (freedom from the cycle of birth and death), have to do with the non-material world (Singh, 2009). *Sanskaras* are accumulated and result in a good life or eternal peace for the soul (*moksha*).

The term *sanskara* is often understood to be merely something like *gunas* (attributes). *Sanskara* is also said to be the basis of culture, which is known in Sanskrit and Hindi as *sanskriti* (K. Singh, 1984). Persons are said to be born with good *sanskaras* (persons are said to be *susanskrit* when they are cultured) or bad ones. From the viewpoint of psychological understanding of Indian culture and religious values as related to travel, we can say that the Indians knew a lot about human psychology. Psychologist Sigmund Freud talked of the subconscious and its role in behaviour and Carl Jung went one step further and connected the individual to culture, the latter of which he called 'the collective unconscious' (Jacobi, 1962; Ricoeui, 1970). Ancient Indians talked about the individual unconscious as connected to the collective unconscious (Ballentine, 1979). In other words, the mind consists of the conscious, the subconscious, and the unconscious, the last of which is connected to your culture (collective unconscious) and your ancestors and which you can access through meditation. This is known as *sanskara*, which people are born with, and *karma*, which you accumulate.

According to Hindu philosophy, it is your *sanskaras* that determine what kind of family you will be born into and your *karmas* decide who *you* ultimately become; because you have the power and ability to choose between good and not-good, or good and bad. This spiritual power can be exercised through slow travel and meditation (which can be done sitting at home or in scenic places where most pilgrim sites in India are usually located) (Singh, 2003) and by taking good care of guests and travellers, subsumed in the Sanskrit sentence 'Atithi devo bhavah' (a guest is a god). Since most yoga tourism in India today is largely about yogic *asanas* and not meditation (as the true purpose of yoga) a new form of yoga 'meditation tourism' could be promoted with great success given the current growing popularity of meditation in the West. Such tourism can also be developed in other countries which have rich histories of spiritual practice.

To explain this form of yoga tourism I refer to the accumulation of *punya* or religious merit or capital that can lead to a good material life but cannot allow you to get *moksha*, which can be done only through spiritually undertaken travel or pilgrimage (Singh, 2004a, 2009). There is also the accumulation of good deeds (like earning a lot of money and giving it *all* away to others, as legendary King Harish Chandra is said to have done, and is also attributed to the founder of Sikhism, Guru Nanak; the same is true for the founder of Buddhism, Siddhartha Gautam Buddha). By such deeds, one accumulates all social, cultural and religious capital but not always spiritual capital, which can only be earned through right prayer and right meditation (i.e. praying and meditating for others and not yourself). That is why yoga tourism, a form of slow travel, has emerged as a distinctive contribution of India, and is being capitalised on by the government.

There is a lot of potential for India to develop yoga tourism further. However, the true meaning of yoga and its philosophy has yet to be delivered to tourists and has not yet attracted good guides or gurus, although there has been a proliferation of so-called 'religious' leaders in India in the past 15 years or so. This can be developed by gurus-turned-leaders, and a few such yogis have made an appearance on the Indian scene in the past five years. For example, the international followers of (late) Swami Yogananda Paramhansa, who have mass meditation sessions in various parts of India, including the scenic mountainous state of Himachal Pradesh. So such tourism does exist in India but needs to be further developed as a form of slow cultural tourism. Such tourism may indeed become India's unique selling point and compete with rising forms such as medical tourism, since meditation and spirituality have therapeutic value (Singh, 2009).

Practical Aspects of Slow Travel Development in India

It may be noted that slow travel, as explained here, does not mean merely travel that is slow in terms of velocity, although bullock carts, mules and horses can be and are used for travel in India. Slow travel is a relative term: use of cars and aeroplanes creates more of the greenhouse gas carbon dioxide than use of buses or trains. Cars, however, allow greater appreciation of landscapes and cultural features along the route to the destination than aeroplanes or trains. However, per capita carbon emission is greater for cars than for buses and trains, hence, along with rural means of transport and journeys undertaken on foot (religious tourism and pilgrimage), buses and trains are essential features of slow travel. Especially appealing are the tourist trains such as the northern and southern 'Palace on Wheels' trains, 'Royal Rajasthan' and some 35 circuits where 'toy' trains and steam engine-driven trains are being used. Such trains would constitute a means of slow travel since they allow tourists to stop along the route and have sufficient time for leisure activities such as sightseeing, meditation, shopping, dining and cultural programmes at hotels. Thus slow travel, as defined earlier, can be good for the environment and for tourists and hosts.

It has been noted that yoga tourism has not been fully exploited by the Government of India, or, at least, not in its sense of meditation tourism. For example, International Yoga Week is celebrated every year from 2–7 February in Rishikesh, Uttarakhand. Though various yogic postures and the value of vegetarianism are explained, little effort has been made to explain why eating meat and eggs are not conducive to meditation. For instance, this author's visit to Rishikesh in 2004 and some interviews undertaken there revealed that the Indian yogic 'gurus' allowed foreigners to eat meat in order to attract them in the initial stages (so that there is no cultural dissonance). Now, though yoga philosophy is not really a Hindu philosophy in the sense that it can be adapted to most religions, including Christianity (Swami Rama, 1975), its essence is rooted in the Hindu way of life and the *karma* philosophy. *Karma* lays emphasis on non-violence and slow food, the latter of which can be defined as food that is absorbed gradually by the body, is low on calories and high in nutrition, does not result in accumulation of fat and does not generate *kaama vasna* or lust. Such food, by definition, is vegetarian. Again, food taken from non-vegetarian sources requires violence, which is not congenial with yogic *karma, gyaan* and *bhakti* yoga, since *karma* yoga says (as the Bible says) that you should not do unto others as you would have others not do unto yourself. *Bhakti* yoga philosophy explains that divine love cannot be attained with violence in your mind (Radha Soami Satsang, 2011: 35).

It may be asked, is all slow travel beneficial? Naturally, slow travel depends like all other forms of travel and tourism on suitable infrastructure and proper promotion. Tourist trains are costly to build and their higher-priced train tickets cannot be afforded by the masses. So, slow travel may appear to be biased towards middle class tourists coming to India. This target market is important since it is wealthier tourists who can more readily afford and prefer flying rather than travelling by train. Moreover, religious journeys undertaken by rural means of transport by the middle classes are also beneficial, but again this depends upon proper promotion rather than infrastructure. In the National Capital Territory of India, some commercial picnic places and farms are creating novel agri-tourism experiences that use camels and bullock carts to enable a slow, rural feel. The success of small tourism businesses indicates how slow travel, if properly planned and underpinned by ecological values, can contribute to sustainable development in the medium term. Implied here is the fact that not all forms of ecotourism (Singh, 2004) and volunteer tourism (Guttentag, 2009, 2011) are beneficial for the environment or generate proper sympathy/empathy and, thereby, are able to account for psychological and ecological tourism carrying capacity (Singh, 2004).

One of the central tenets for a philosophy of slow travel is care for the environment. Traditional Indian slow travel was informed by these values, but modern-day forms of tourism in the same places are becoming hedonistic. By becoming informed of this philosophy in tourism promotion in India (and other countries), slow travel and tourism can become a larger reality.

New forms of slow travel have emerged in India, such as ecotourism, volunteer tourism, village tourism and pro-poor tourism, but, like slow travel in the West, have yet to generate philosophical roots. As a result, it is in the interest of the Government of India to draw upon the ancient philosophy of travel and not seek over-commercialisation of tourism in the short run, but, rather, find it in reinforcing well-thought-out forms of slow travel, such as pilgrimage, religious and cultural tourism, as well as growing areas such as agri-tourism. Some other examples are Camel Safaris and the Palace on Wheels railway package in Rajasthan, rural and village tourism in Goa and Maharashtra, and Ayurveda (Hindu medicinal system) tourism in Kerala; all are very popular. To some extent, the Indian government has realised this and promoted or facilitated slow travel such as pilgrimage and ecotourism. This has been done by constructing concrete and easy-to-use pathways in remote areas such as the pilgrimage sites of Kedarnath and Har-ki-paidi (Haridwar), and eco-treks such as the difficult Har-ki-doon mountain trek, all in the Garhwal region in the hill state of Uttarakhand. Another pilgrim route travelled by Hindus and Sikhs in Jammu and Kashmir state, Vaishnodevi, also has had a concrete path built right up to the shrine for some years now and is perhaps the most popular slow travel route in north India, with one million visits in 2003 (Singh, 2004). However, progress in making travellers more eco-conscious has been slow, with visible solid waste and water pollution being common at these sites.

Unlike the city of Chandigarh (close to the Himachal Pradesh Shiwalik mountain range), where use of plastic and polythene bags has been banned and a fine imposed on violators, Uttarakhand and Himachal Pradesh states have yet to curb such pollution. Routes like the one to Vaishnodevi also show that the character of slow travel has not changed much, and only the accent has changed, with pilgrims deriving more recreation from their journey than in the 1950s and 1960s, when many mountain roads had just been built and public transport was limited, causing tension en route. But there is at least one example of a popular pilgrimage site where the government, in collusion with private contractors, has attempted to change the character of slow travel. In Kedarnath, Uttarakhand, one can now hire a helicopter from Dehradun in the foothills and complete the to-and-fro journey in a day. While it may be argued that this has been done to promote 'fast' tourism, the deeply cultural and spiritual experience that is 'sold' to tourists has been compromised.

In conclusion, I refer to the case of Rishikesh in the Uttarakhand foot-hills. Although there are no reliable statistics of foreign tourist visits to Rishikesh, it is estimated that about 30% of the foreign tourists who visit India visit the Uttarakhand Himalayas, whose gateway is Rishikesh. Foreign tourist arrivals in India in 2010 stood at approximately 4.92 million (Businessworld, 2010a). Not all of these, however, visit the Uttarakhand area in February when the International Yoga Festival is held. Overall visits to Uttarakhand stood at 20,546,323 in 2010 (Businessworld, 2010b). A visit by the author in 2004 revealed that foreign tourists are not clear about the Indian idea of slow food, such as germinated grains and vegetables instead of non-vegetarian food and confectionery; shops selling the latter are to be found in Rishikesh to attract foreign tourists. Also, some tourists were overheard mentioning that sexual overtures by fellow 'yogis' were not uncommon. Similar ideas about yoga and those who come to practice meditation at Osho Ashram in Pune in Maharashtra have largely given foreign tourists the idea that spirituality is not divorced from sex, whereas from the Indian viewpoint they are antipodal. Thus, slow travel and spirituality, as understood by traditional Indians, have been misinterpreted by Westerners travelling to India or other countries where yoga and meditation have become popular.

Conclusion

India is the home of a variety of religious and spiritual philosophies. However, these philosophies are frequently esoteric and are kept secret by the custodians of religion – *pundits* in the case of Hinduism. Moreover, even the custodians of religion often do not understand the deep meanings of these texts, since, at least among Hindus, religious functionaries get stuck with rituals and cannot decipher the multiple meanings, as happens in Sanskrit texts. Travelling drama troupes such as *nautankis* and *jatra* performed the function of keeping this culture consistent and spread it among the masses over the centuries. As a result, a complex civilisation and culture like India's has remained wedded to those social processes that keep its societal structure and essential ecological processes alive and well, including the process of slow travel.

The definition of slow travel taken in this chapter has some conceptual overlaps with, but remains distinct from, wellness tourism as variously defined. These conceptual overlaps, however, establish the veracity of slow travel as a conceptual tool in the scheme of tourism studies. For Mueller and Kaufman (2001), wellness is broader than health, and includes physical fitness, body and beauty care, nutrition and diet, relaxation, rest and meditation, and mental activity, all built around taking responsibility for ones health and embedded in positive social contacts and environmental sensitivity. They include meditation and mental activity and education (Mueller & Kaufman, 2001: 7), which is commensurate with slow travel definitions. Environmental sensitivity is also common to slow travel and wellness tourism. Wellness tourism is considered a broad term by other authors as well, such as Laing and Weiler (2008) and Smith and Puczko (2009). The latter identify and describe a spectrum of wellness tourism activities ranging from physical healing through rest and relaxation and spiritual pursuits. According to Smith and Puczko (2009), each point in this spectrum is associated with specific activities and types of tourism products, with spiritual wellness tourism being associated with pilgrimages and yoga centres. Similarly,

Smith *et al.* (2010: 90) include yoga, meditation and spiritual activities at retreats and ashrams as part of wellness tourism. A similar conceptual overlap is found in Sheldon and Bushell's (2009: 11) definition of wellness tourism, who term it 'a holistic mode of travel that integrates a quest for physical health, beauty or longevity, and/or a heightening of *consciousness* or spiritual awareness, and a connection with *community, nature* or the *divine* mystery' (emphases introduced by this author). The difference between slow travel and wellness tourism here is that in the former the modes of transport used and their effects on the environment are crucial. However, just like wellness tourism, slow travel and tourism is a broad concept that includes a wide spectrum of activities and tourism products.

To conclude the theoretical analysis, it may be asked: what is the social function or purpose of slow travel, apart from what has been already outlined? In answer to that question it may be said that the social function of all tourism, but especially slow tourism, is to keep people from different societies and cultures in respectful contact with each other. The social function of slow travel is even more important for human society than other forms of tourism. This is because it is only through slow travel that other cultures can be properly understood, an essential process for globalisation and localisation. This can be formulated into an anthropological theory of slow tourism, as follows:

> Cultures exist in relation to each other, and slow travel and tourism are means of spreading, electing between, and imbibing non-material cultures largely through material means and ends, leading, when properly done, to cultural relativity – an anthropological value that can promote peace and understanding by downplaying the tendency of people to consider their own culture as superior – and stability of essential ecological processes.

To support and conclude this analysis, it should be emphasised that the process of imbibing and electing between cultures and values of another social system has to be a slow process. Culture cannot be absorbed in a day and hence we have seen the emergence of the now well-known category of 'anthropologist tourists' (Gibson & Yiannakis, 2002).

References

Ardener, E. (1972) Introduction. In E. Ardener (ed.) *Social Anthropology and Language* (ASA Monograph 10, pp. 1–32). London: Tavistock.

Ballentine, R. (ed.) (1979) *Yoga and Psychotherapy.* Honesdale, PA: Himalayan International Institute of Yoga Science and Philosophy.

Basham, A.L. (1978) *The Wonder That Was India.* New Delhi: Macmillan.

Bose, N.K. (1967) *Culture and Society in India.* Bombay: Asia Publishing House.

Buckley, R. (2003) *Case Studies in Ecotourism.* Oxford: CABI Publishing.

Businessworld (2010a) Foreign tourist arrivals in India (January to November) – Online document, accessed 9 December 2010. http://www.businessworld.in

Businessworld (2010b) Number of tourist visitors statewise – Online document, accessed 5 May 2010. http://www.businessworld.in

Fennell, D. (1999) *Ecotourism: An Introduction*. London: Routledge.

Gibson, H. and Yiannakis, A. (2002) Tourist roles: Needs and life course. *Annals of Tourism Research* 29 (2), 358–383.

Graburn, N.H. (1983) Editorial: The anthropology of tourism. *Annals of Tourism Research* 10 (1), 9–33.

Graburn, N.H. (1989) Tourism: The sacred journey. In V. Smith (ed.) *Hosts and Guests: The Anthropology of Tourism* (2nd edn) (pp. 17–31). Philadelphia: University of Pennsylvania Press.

Guttentag, D.A. (2009) The possible negative impacts of volunteer tourism. *International Journal of Tourism Research* 11 (6), 537–551.

Guttentag, D.A. (2011) Volunteer tourism: As good as it seems? *Tourism Recreation Research* 36 (1), 69 74.

Heitmann, S., Robinson, P. and Povey, G. (2011) Slow food, slow cities and slow tourism. In P. Robinson, S. Heitmann and P. Dieke (eds) *Research Themes for Tourism* (pp. 114–127) Wallingford: CAB International.

Jacobi, J. (1962) *The Psychology of C.G. Jung: An Introduction*. New Haven, CT: Yale University Press.

Kaur, J. (1985) *Himalayan Pilgrimages and the New Tourism*. New Delhi: Himalayan Books.

Kobasic, A. (1996) Level and dissemination of academic findings about tourism. *Turizam* 44 (7–8), 169–181.

Laing, J. and Weiler, B. (2008) Mind, body and spirit: Health and wellness tourism in Asia. In J. Cochrane (ed.) *Asian Tourism: Growth and Change* (pp. 379–390). Oxford: Elsevier.

Misra, L.K. (ed.) (1976) *The Art and Science of Meditation*. Glenview, IL: Himalayan International Institute of Yoga Science and Philosophy.

Mueller, H. and Kaufman, E. (2001) Wellness tourism. *Journal of Vacation Marketing* 7 (1), 5–17.

Nadel, S. (1942) *Foundations of Social Anthropology*. Oxford: Clarendon Press.

Nash, D. (1989) Tourism as a form of imperialism. In V.L. Smith (ed.) *Hosts and Guests: The Anthropology of Tourism* (pp. 33–47). Philadelphia: University of Pennsylvania Press.

Puri, J.R. (1982) *Guru Nanak: His Mystic Teachings*. New Delhi: Radha Soami Satsang Beas, Punjab.

Radha Soami Satsang (2011) The divine path. *Spiritual Link* 7 (1), 1–38.

Ricoeui, P. (1970) *Freud and Philosophy*. New Haven, CT: Yale University Press.

Sharma, R.S. (1979) *Ancient India*. New Delhi: NCERT.

Sharma, R.S. (1987) *Society and Culture in Ancient India*. New Delhi: Macmillan.

Sheldon, P. and Bushell, R. (2009) Introduction to wellness and tourism. In R. Bushell and P.J. Sheldon (eds) *Wellness and Tourism: Mind, Body, Spirit, Place* (pp. 3–19). New York: Cognizant.

Singh, K. (1984) *Indian Society*. Lucknow: Kitab Mahal.

Singh, S. (1986) The ecological impact of tourism in the Garhwal Himalayas. Proceedings of the national workshop on the Future of Tourism in India and its Implications, Bangalore, India, May 1986 (mimeo).

Singh, S. (2002) Managing the impact of tourist and pilgrim mobility in the Indian Himalayas. *Revue de Géographie Alpine* 90 (1), 25–34.

Singh, S. (2003) Travel and aspects of societal structure: A comparison of India and the United States. *Current Issues in Tourism* 6 (3), 209–234.

Singh, S. (2004a) Religion, heritage and travel: Case references from the Indian Himalayas. *Current Issues in Tourism* 7 (1), 44–65.

Singh, S. (2004b) *Shades of Green: Ecotourism for Sustainability.* New Delhi: The Energy and Resources Institute Press.

Singh, S. (2005) Volunteer tourism in India: Square pegs for round holes? *Tourism Development Journal* 2&3 (1), 28–31.

Singh, S. (2009) Spirituality and tourism: An anthropologist's view. *Tourism Recreation Research* 34 (2), 143–155.

Singh, S. (2011a) Religious tourism, spirituality and peace: Philosophical and practical aspects. In *Religious Tourism in Asia and the Pacific* (pp. 15–21). Madrid: UNWTO.

Singh, S. (2011b) The tourism area 'life cycle': A clarification. *Annals of Tourism Research.* 38 (3), 1185–1187.

Smith, M., Macleod, N. and Robertson, M. (2010) *Key Concepts in Tourism Studies.* London: Sage.

Smith, M. and Puczko, L. (2009) *Health and Wellness Tourism.* Amsterdam: Butterworth-Heinemann.

Swami Ajaya, Dave, J., Nuernberger, P., Bates, C. and Ballentine, R. (eds) (1979). *Therapeutic Value of Yoga.* Honesdale, PA: Himalayan International Institute of Yoga Science and Philosophy.

Swami Rama (1975) *Meditation in Christianity.* Glenview, IL: Himalayan International Institute of Yoga Science and Philosophy.

Times News Network (2010) More pilgrims than tourists in India. *The Times of India,* 27 September, p. 1.

Timothy, D. and Olsen, D. (eds) (2006) *Tourism, Religion and Spiritual Journeys.* London: Routledge.

Wearing, S. (2001) *Volunteer Tourism: Experiences that Make a Difference.* New York: CABI Publishing.

Wolf, E. (1966) *Peasants.* New York: Pearson.

17 Reflecting Upon Slow Travel and Tourism Experiences

Kevin Markwell, Simone Fullagar and Erica Wilson

In his ground-breaking book, *The Holiday Maker: Understanding the Impact of Leisure and Travel*, first published in German in 1984, Swiss academic Jost Krippendorf set out what he called his 'credo for a new harmony' (1987: 10). Concerned about the impacts that modern mass tourism had inflicted on destinations, communities and also on tourists themselves, Krippendorf asked 'Must we in the future, in order to get on, run twice as fast as before...? Shouldn't we instead take the foot off the accelerator if we want to win the race?' Anticipating the movement towards alternative tourisms that embraced ecological sustainability and fair exchange between local and tourist by about a decade, Krippendorf's work can also be read as an argument for forms of tourism practice that encourage a more genuine and sustainable engagement with places, cultures, natures and peoples. Not only did he anticipate the emergence of sustainable tourism, but he also anticipated the emergence of the slow movement, and slow travel and tourism in particular.

Following the publication of *The Holiday Makers*, an extensive and sustained body of work has been published that has reported on, critically examined, or advocated for, alternative tourism practices that seek to minimise social, cultural and environmental impacts while at the same time creating a greater level of quality experience for the local and the tourist. Beginning with ecotourism, which first entered academic discourse as a named form of tourism in the late 1980s, the nomenclature has included 'alternative tourism', 'sustainable tourism', 'new tourism' 'responsible tourism' and 'pro-poor tourism'. These tourisms were not seen only as new forms of 'special interest tourism', although, of course, each does have a place within the special interest or niche tourism umbrella, but also as forms of tourism that challenged the dominant model of mass tourism that developed following the ending of World War II. The growth in mass tourism reflected the rapid rate of economic growth enjoyed by many developed economies, advances in transportation technology, ready availability of oil and an optimistic desire by the affluent middle classes to acquire cultural capital through travel. The work of researchers, community advocates, environmentalists and policy-makers has, however, uncovered the inherent unsustainable basis for conventional mass tourism.

Running parallel to these concerns about mass tourism has been the development of a growing slow movement that rejects what might be called the hegemony of speed (see McQuire, 2000; Virilio, 1986) and its concomitant affects on the quality of social, cultural and environmental relations. The Slow Food movement that began with Carlo Petrini's organised protests against the imposition of a McDonald's fast food restaurant in his home town has developed into an all-embracing slow living movement. Slowness works as a metaphor that brings into question the cult of speed and embraces an approach to life that values time in terms of relationships between people and place. Apart from the work of on-the-ground advocates, the movement's success in America and the UK is evident in the huge popularity of Carl Honoré's (2005) social commentary *In Praise of Slow*. Honoré argues that we are living in an era where speed has assumed greater importance than in the whole of human history. Elite athletes compete within a hundredth of a second with each other as the technologies to measure such miniscule differences in speed have been developed. Answers to once complicated and difficult questions are now a computer search engine away that can deliver us answers in a nanosecond. A recent survey in Britain found that 60% of teenagers and 37% of adults are 'highly addicted' to their smart phones, never wanting to part from them for an instant – even when in the bathroom or in bed – in case they miss something important (Halliday, 2011). Tabloid newspaper reports inform us of the latest incidents of road rage and queue rage which reflect our growing impatience for delays of any sort. Speed is our god. It is no wonder that the growth of counter movements like 'International Take Back Your Time Day' have arisen to question whether a fast life is a life well lived.

Along with slow food a number of other slow entities and movements have emerged, such as the slow city, slow money, slow parenting, slow sex, slow work and of course, the focus of this book, slow travel and tourism. The project of slow travel/tourism encompasses a philosophical position that resists the homogenising forces of globalisation and the notion of tourism as a commodified experience of mobility and instead offers an alternative vision that celebrates the local; small-scale travel utilising transport modalities that minimise the impacts on the environment and facilitate a closer and more genuine connection with local people. Slow tourism foregrounds the notion of convivial hospitality as being a crucial element in the slow tourism experience and as such strengthens the relations between local and visitor (Conway & Timms, 2010). While there has been an explosion of slow tourism websites and blogs over the past few years, and 'products' are now available that provide for the slow tourist, there has, to date, been little critical attention paid to the concept by academics.

Perhaps the first author to argue for a slow approach to tourism within an academic context was Rafael Matos, an economic geographer, who wrote a chapter titled 'Can slow tourism bring new life to alpine regions' published in the book *The Tourism and Leisure Industry: Shaping the Future*, published in 2004 (Weiermair & Mathies, 2004). For Matos, slow tourism was founded on two principles: 'taking time' and 'attachment to place' (Matos, 2004: 100) and his chapter rehearsed the arguments for slow travel and tourism that underpin much of the more recent work in this area. Putting into practice his theories, he even suggested that a form of hotel that he termed 'slowtel' could develop that embraced and celebrated the slow ideal. A focus on slow travel as a response to the crisis of climate change and growing carbon pollution has been at the heart of a number of recent publications by Janet Dickinson and Les Lumsdon, including their book, *Slow Travel and Tourism* (2010). C. Michael Hall has also contributed to the literature through his analysis of the slow movement, and, in particular, slow food, in relation to sustainable tourism (see Hall, 2006). Jennie Germann Molz (2009: 283), in her examination of the relationships between tourism, pace and modernity, found that acceleration was associated with concepts such as 'success', 'beauty' and 'freedom' while slowness was seen to be associated with 'incorrect' or 'undesirable' ways of travelling. Finally, Conway and Timms (2010) argued that slow tourism is a sustainable alternative to mass tourism in the Caribbean islands and provided a series of case studies of what they see as examples of slow tourism that are already operating, with some success, in that particular region.

The genesis of this book, then, is situated in this emerging literature on slow travel and tourism. However, our aim has not been to provide a definitive account of slow travel, as that would be antithetical to the mobility of meaning which characterises different cultural practices, interpretations and methodological explorations. Instead, the diversity of contributions in this collection explored different dimensions of slow that may continue to open up new ways of problematising and examining emergent tourism practices and systems. As we hinted at in our introductory chapter, we came to the topic with a mixture of intellectual curiosity as well as a personal interest in the slow ethos which is reflected in various ways in our everyday lives. For one of the editors, it was his reflections on his and his partner's experiences as a host property listed on HelpX's website, (an organisation very similar to Willing Workers on Organic Farms) that led him to seriously consider the concept of slow travel and tourism. Over the past couple of years, he and his partner have hosted nearly 30 young travellers who have spent between four days and five weeks staying with them, exchanging their labour for accommodation in their house and inclusion in the routines and rhythms of the household. None of these travellers has mentioned the term 'slow tourism' of course, but many of them, when asked why they decided to travel this way, have identified the desire to stay somewhere for a longer period of time, to get to know a smaller place in the country, and 'close to nature', and to gain a better understanding of what it means to live in Australia. For some, these desires, understandably, intersected with a need to also save money.

The experiences have been mutually beneficial and both 'host' and 'guest' have enjoyed this form of cultural exchange framed, in part, in a context of hospitality. Yet, the paradoxes and contradictions, inherent no doubt, in any form of alternative tourism, have become increasingly clear. This 'slow' component of their overall travel experience (for some it will be the only one, for others it will be one among many) is inevitably situated within a more ortho-dox or conventional 'fast' tourism. They have all been international tourists who have boarded jet aircraft to travel to Australia as quickly and as inex-pensively as they could. Most have travelled within Australia by coach, but some took flights if they could, while some others bought their own vehicle to give them greater flexibility and to save on travel costs which could be more easily shared through joint ownership of a vehicle. Probably the most striking thing, however, has been the inclusion of a laptop computer or e-tablet amongst their luggage. Only a handful of travellers have not trav-elled with one.

It is ironic that these computers are the means to retain the connection to the fast world from which these travellers are seeking a temporary escape. Using instantaneous messaging or Skype (email is deemed too slow), they maintain contact with their family and friends via an exchange of digitised data that facilitates communication in nanoseconds from rural Australia to towns and cities in the northern hemisphere. Indeed, it could be argued, fol-lowing Poon (1994), that this new practice of slow tourism is in part depen-dent on the super-fast communication technologies. The internet enables consumers to learn about the slow movement, to locate slow tourism prod-ucts and to interact with other slow tourists through blogs. Space-time com-pression may still be a necessary, if 'backstage', component of slow travel and tourism.

The editors approached the concept of this book cognisant of the incon-sistencies, paradoxes and contradictions as well as the opportunities for a more equitable and sustainable tourism that characterise this different way of doing tourism. Indeed, it is the paradoxes and inconsistencies that demand the critical attention of scholars who can interrogate slow travel and tourism from a range of disciplinary perspectives. The 'elephant in the room' for any discussion of the ecological sustainability credentials of ecotourism, sustain-able tourism or slow tourism, is of course, the energy source used to move the tourists from origin to destination and back again. It is an inconvenient truth that the trip commences upon departure from the origin, and that usu-ally involves the use of non-renewable energy that emits carbon into the atmosphere. As Fullagar outlined in her chapter, some of the cyclists she studied acknowledged the environmental impact that they produced as they travelled, sometimes from interstate, sometimes internationally, to begin the cycling event. Hall, Parasecoli and de Abreu e Lima, pointed out in their respective chapters, that while slow food events and destinations serve as desirable attractions for slow travellers, what proportion of these travellers are using low-technology modes of transport to travel to these places to experience slow food?

Negotiating a 'slow life' that embraces slow travel and tourism as lived experience within an otherwise 'fast world' is difficult and complex and bound to draw attention to inconsistencies of practice and contradictions in ethos. The long-term wanderers who were the subject of Tiyce and Wilson's chapter are all dependent on their own private motor vehicle (in this case, motorhomes) to move them (slowly) around Australia, while the hitch-hikers of O'Regan's chapter are dependent on the motor vehicles of others to transport them. Lipman and Murphy's WWOOFers also highlight the inconsistency of slowing down, becoming more closely connected or even integrated into local communities, while still usually being dependent on non-renewable energy to reach the destination in the first place. The matter of mode of transport was taken up by several other authors in this book, who examined pilgrimages undertaken through the practice of walking (Howard, Singh) or through boating on the canals of England (Fallon). Each of these chapters reminds us of the significance of our choice of energy that powers our slow tourist experience.

The opportunities to create a new form of tourism practice that potentially reduces our environmental and social impacts, while simultaneously enhancing our individual experiences, were the focus of a number of chapters in the book. Moore argues for an ethic of travel that creates a deeper sense of meaning through a rejection of overly-materialist values and an understanding of well-being that connects individuals, communities and the environment. A rejection of materialism and of capitalism's propensity to commoditise relations among people and between communities and the environment underpins Wearing, Wearing and McDonald's chapter.

These themes relating to travel, time, sustainability, connectivity and identity are then taken up and explored in a number of case studies of particular destinations. de la Barre's study explored the mobilisation of discourses around 'slow time' to market a destination that might otherwise have been faced with access and infrastructure challenges. However, by taking this approach the marketing strategies were seen to be implicated in the process of othering the Indigenous peoples of the region. The opportunities for rural development in Japan were the focus of the chapter by Murayama and Parker, who assert that such tourism does have considerable potential for economic and social development, even in a nation that has been emblematic of 'fast living'. A nascent desire to escape fast living on the part of an increasing number of urban dwellers is seen to be fuelling an interest in slow modalities. Gibson, Pratt and Movono's example of the 'Tribewanted' project in Fiji is an interesting case of slow tourism embracing volunteerism and sustainability, supported in part by an on-going cyber community of past and potential volunteers. Finally, the idea of lifestyle entrepreneurship is explored in Groenendaal's examination of Dutch people emigrating to France to open B&B accommodation. She found that these lifestyle entrepreneurs gained considerably by the fact that they themselves were living away from their homeland and that they played an important role in helping to shape slow tourism as an economic practice.

The chapters in this book, ranging from philosophical and theoretical explorations to empirical case studies, contribute to scholarship into slow travel and tourism and help to extend our understanding of its potential for creating a more sustainable and equitable tourism practice. But many questions still remain and these form the basis of future investigations. The extent to which slow tourism is regarded as just another form of special interest or 'adjectival' tourism or whether it is able to help transform tourism, is open to speculation. Is 'slow time' yet another 'foreign country', to invoke David Lowenthal (1985), to be explored and colonised by our insatiable quest for novelty? Is it simply the 'next hot thing' as Parasecoli and de Abreu e Lima (in this volume) ask? Or, alternatively, is slow the new 'small', in the words of E.F. Schumacher in his popular 1973 book, *Small is Beautiful: Economics as if People Mattered*? Could slow tourism lead to a fundamental transformation in tourism? Will the drivers of such change paradoxically include the limits of fast growth (peak oil for example), the rise of glocalisation, where travelling closer to home is re-valued, or a health-related scenario in which slow travel arises from a desire to recoup from intensified work and economic pressures? And yet, how could this possibly happen given the rapidly growing consumer markets in China and India which will crave the same kinds of travel opportunities that developed economies have been enjoying for the past half century or so? Will slow tourism become polarised into either a boutique form of travel experienced by those who have the 'luxury' of travelling slow or a mobility decision (such as the bicycle) that arises from economic constraints and the challenge of frugality?

In keeping with the case studies in this book, there is a need for a range of studies that document the emergence of slow travel and tourism as an alternative form of tourism within the broader context of the slow movement. A case studies approach could also enable useful examinations of approaches to 'product development' and marketing from a slow tourism perspective. Ideally, these case studies would go beyond the dominant Anglo-American centrism of much tourism scholarship and include examples from a range of nations from both developed and developing economies. From the demand side, there is much research to be done examining the propensity for consumers (from different socio-economic backgrounds) to embrace the ideals and philosophy of slow tourism. How might slow travel be experienced as a form of alternative hedonism (Soper, 2008) that evokes pleasure through practices of 'treading lightly' rather than status-oriented consumerism? What are the reasons that lead to some people accepting the proposition that slow tourism brings with it: that conventional ways of doing tourism are inherently unsustainable and inequitable? Why are others so resistant to change? These questions are about travel and tourism at one level, and at another level they are about our very way of living within and relating to the world in the search of the 'good life', 'wellbeing' and 'sustainability'. Importantly, the metaphor of slow resonates in a number of ways to open up a discursive space that may continue to fuel the popular tourism industry and academic imagination.

There may be no easy or, dare we say, 'quick' answers to these questions. As we stipulated in the introduction to this book, we did not set out to identify one easily recognisable definition for slow travel and tourism. Rather, we wished to open up the concept of 'slow tourism' to critical analysis and to encourage further research in order to gain a more nuanced understanding. Ultimately, it is our hope that this book will contribute to ongoing debates concerning slow tourism and the search for alternative mobilities.

References

Conway, D. and Timms, B.F. (2010) Re-branding alternative tourism in the Caribbean: The case for 'slow tourism'. *Tourism and Hospitality Research* 10 (4), 329–344

Dickinson, J. and Lumsdon, L. (2010) *Slow Travel and Tourism*. London: Earthscan.

Germann Molz, J. (2009) Representing pace in tourism mobilities: Staycations, Slow Travel and the Amazing Race. *Journal of Tourism and Cultural Change* 7 (4), 270–286.

Hall, C.M. (2006) Introduction: Culinary tourism and regional development: From slow food to slow tourism? *Tourism Review International* 9 (4), 303–305.

Halliday, J. (2011) Facebook and Twitter fuel iPhone and Blackberry addiction, says Ofcom, *The Guardian*, 4 August, accessed 25 September 2011. http://www.guardian.co.uk/technology/2011/aug/04/facebook-twitter-iphone-blackberry-addiction-ofcom

Honoré, C. (2005) *In Praise of Slow: How a Worldwide Movement is Challenging the Cult of Speed*. London: Orion.

Krippendorf, J. (1987) *The Holiday Makers: Understanding the Impact of Leisure and Travel*. Oxford: Heinemann Professional Publishing.

Lowenthal, D. (1985) *The Past is a Foreign Country*. Cambridge: Cambridge University Press.

Matos, R. (2004) Can slow tourism bring new life to alpine regions? In K. Weiermair and C. Mathies (eds) *The Tourism and Leisure Industry, Shaping the Future* (pp. 93–104). Bighamton, NY: The Haworth Hospitality Press.

McQuire, S. (2000) Blinded by the (speed of) light. In J. Armitage (ed.) *Paul Virilio: From Modernism to Hypermodernism and Beyond* (pp. 143–160). London: Sage.

Poon, A. (1994) The 'new tourism' revolution. *Tourism Management* 15 (2), 91–92.

Schumacher, E.F. (1973) *Small is Beautiful, Economics as if People Mattered*. New York: Harper and Row.

Soper, K. (2008) Alternative hedonism, cultural theory and the role of aesthetic revisioning. *Cultural Studies* 22 (5), 567–587.

Virilio, P. (1986) *Speed and Politics, an Esasy on Dromology*. New York: Semiotexte.

Weiermair, K. and Mathies, C. (eds) (2004) *The Tourism and Leisure Industry, Shaping the Future* (pp. 93–104). Bighamton, NY: The Haworth Hospitality Press.